# TH HUMAN POPULATION TSUNAMI

## How it could be managed

Martin Jacoby

## The Author

Martin Jacoby is a lifelong naturalist, including 25 years as a professional. After reading Zoology at Oxford, he spent much of his working life teaching. He has led about 150 journeys in many parts of the world to look at the environment and explain how it works. His main relaxation is managing a little bit of ancient woodland.

## Acknowledgements

Henry Bewer, Robin Bruce, Antony Hichens, Wolfgang Hickel, George Low, Richard Luce, Inka Wilden and Mike Wood were all kind enough to read the first edition of this book. Frank Brierley, Tom Fisher, Christian Michel, Bill Radcliffe, Fauzia Rahman, Jeff Spence and Peter Townsend read drafts of an essay abstracted from it, and Sonia Cutler edited this edition with consummate thoroughness and patience. I am deeply indebted to them all for so generously giving their time, unstinting help and constructive criticism. Inevitably errors of fact and flaws in my arguments remain, but they are entirely mine.

## Copyright



Self-published: September 2013 through CreateSpace
Second printing: November 2013 with a new Introduction and minor corrections.
Second edition: October 2015, substantially revised.

ISBN 978-1492282686

**Cover design:** Inka Wilden

*[Endnote numbers refer to references. Footnote numbers are underlined.]*

# CONTENTS

# PREFACE

I wrote this book to clarify my own lines of thought. I did not write it with hate or despair, but as free from emotion as I could, though occasionally my exasperation shows through. Nor do any of the contentious recommendations I make towards the end represent any desire of mine; indeed, I earnestly wish the opposite of them. They merely seem to be logically the least horrible ways of managing the humanitarian disaster that is upon us. We are already (summer 2015) seeing the tiny foreshocks of what is to come in the mass migration of people into Europe – apart from the almost perpetual wars in deserts that were once fertile lands.

*The Human Population Tsunami* is not a comfortable read; it should not be judged by some emotional *frisson* it might elicit but by accuracy of reference and logicality of thought. The book includes the basic information that a reader without a science background will need to follow my arguments; accordingly, I make no apology for starting with the fundamentals of each topic I treat, including some elementary chemistry. We live in a scientific age, and only those who understand the principles behind the first two lines of the periodic table of chemical elements can realistically call themselves either educated or cultured.

Likewise, understanding the mechanisms of both evolution and ecology is as essential as elementary physics and chemistry to all thinking human beings. These topics, together with some knowledge of the steps our ancestors took in their evolution into our modern selves, are prerequisites to being able to judge effectively how to manage our collective future. Rational understanding forms the only common ground between people of all religions and convictions. Let us meet on it.

Martin Jacoby, Stawley, October 2015

# CHAPTER SUMMARIES

## Chapter 1: The Problem

In July 2015, our population was estimated to exceed 7.3 billion and increasing by 80 million each year. The United Nations (UN) Universal Declaration of Human Rights states that everyone has the right to a reasonable standard of living. This is impossible to achieve while we are reproducing exponentially because the Earth and its resources are finite.

At least 1 billion of us will die in the next 35 years. Either most of us will die full of years and under the rule of law or messily, many while still young, when the concept of property breaks down. Where reality comes to rest between these two extreme scenarios depends on how we answer the question, 'Should we or should we not try to meet the UN's expectations?'

## Chapter 2: Thinking About Answers

The question that ends Chapter 1 is so important that its answer deserves all our intellectual skills, beginning with understanding how we think. With a warning about the reliability of what we call 'truth', we trace the ancestral origins of the mental models we put up against managing the human population tsunami. The beginning of time is as good a point as any to begin telling the story of our evolution.

## Chapter 3: From Cosmology to Cells, Competition & Cooperation

We trace the likely evolution of incandescent plasma that followed the Big Bang into atoms and then molecules. Some molecules compete in collisions that cause chemical change, while others, such as enzymes, cooperate to achieve those changes. Autocatalytic chain reactions evolved from a balance between competition and cooperation, and led to replication becoming an end in itself. Further cooperation, such as between enzymes and energy sources, or nucleic acids and proteins, and then including lipids, led to the evolution of cells.

## Chapter 4: Evolution: the Mechanism & the Story

The mechanism of evolution involves four indispensable requisites: multiplication with inheritance, variation and selection. Evolution's mechanism is distinct from the story of the physical and behavioural changes that resulted from it. The history of Darwin's and Wallace's insight, and the often-neglected contribution of Lamarck, give a background to that story.

The origin of life involved a highly complicated cell, rather than just DNA; and that cell came about by symbiosis between different kinds of living organisms. The difficulty of the early steps in our evolution and the speed of later ones are reflected in a calendar in which we condense Earth's history into one year. Using this scale, the dinosaurs appear on 15 December, and our written history begins 5 minutes before the end of that year.

## Chapter 5: Primates & Man, Tools & Brains

Following, as best we can, the evolution of human behaviour reveals the antiquity of many mindsets we think are unique to us, or at least sophisticated: certainly the use of tools is widespread among other species. Many of these mindsets affect our capacity to cope with overpopulation. The evolution of consciousness, self-consciousness, other-consciousness and empathy paved the way for our ultra-cooperativeness, which in turn led to the looming ecological crisis.

## Chapter 6: Fire & Language, Farming & War

About 200,000 years ago, we evolved a supremely adaptable body; then we were reduced to a few thousand individuals with massive loss of genetic diversity. These two features, combined with language and a rapidly expanding brain, began our transformation into a behavioural, rather than a genetic, species.

We spread out all over the world and fed on large mammals, causing the extinction of many species.

Our rising population and the loss of this food source led to conflict between groups of people, with victory depending on the number of combatants. Farming allowed higher concentrations of people, leading to villages, towns and cities, and the evolution of armies, religions and warfare.

## Chapter 7: Villages & Cities, Religion & Taboos

Villages, in which everyone knew everyone else, evolved to protect growing and stored food from theft. The evolution of cities and individual anonymity allowed cheating and crime. Larger numbers of people led to increasing diversity and organisation, including specialised fighters, who formed armies. The superstitions of fighters evolved into organised religions with both practical and mystical taboos. Breaking taboos was a major way by which the understanding of reality has advanced. Three major steps were demolishing of geocentrism by Copernicus, divine creation by Darwin and religious control of morality by Hamilton.

## Chapter 8: The Special Case of Islam

In this chapter, we look at the central tenets of Islam, one of which is that the life of Muhammad is a model for Muslim men. Thinking about that life from the perspective of naturalism, it is clear that Islam's underlying drive is to outbreed other ways of life, which is why it is discussed in this book.

## Chapter 9: Genes & 'Thenes', Idenes & Individuals

We look at the dual nature of genes, things and ideas, and see that this approach can also be applied to us. Since the materials that make up our bodies are continually changing, the essence of our individuality is the genetic and behavioural information that orders how those materials are assembled and behave. But that information is also changing; this means that each of us is the temporary interaction between two streams: one of materials that flow through us, and the other of information. The tiny stream of information that interests us individually between conception and death is the ultimate subject of natural selection. Hence, our immortality is not the preservation of some stage in our physical existence, but the genetic and behavioural information we pass on to others during our lives.

## Chapter 10: Ecology & Conurbations

An ecosystem is any part of the universe we define in space and time. Energy flows through it, and materials circulate within it, or may be imported and exported. The Earth, as an ecosystem, has practically no import or export. Cities are ecosystems that depend almost entirely on import and export. When the energy for that transport fails, civil life in them disintegrates. This has happened in at least nine cases during recent history. There is no clear evidence that modern economies will escape the same fate.

## Chapter 11: Money

We trace the origin and evolution of money and identify deep inconsistencies in its uses. Some of these were evident in the financial crisis that began in 2007 and is still ongoing in 2015.

As soon as coins cease to *be* the value of the metal in them, they become tokens for that value, which can be redeemed in another place and at another time. Therefore, money, in whatever form, cannot avoid being a form of borrowing. Borrowing depends on confidence that the lender will be repaid in the future. Growing populations and growing economies increase confidence. Ultimately, such confidence lies in the myth of the Earth's infinite resources; and that is the basic standard behind modern money.

Now that the myth is being exposed, we need a new standard that relates economic activity to the Earth's finite resources, particularly renewable energy. I thereby propose a new global currency based on a unit of reduced carbon.

## Chapter 12: Biophilia & Health, Lekking & Birth Control

The connection we feel with all life increases with the percentage of genes we share, so it is greatest in our nearest relatives. In a highly social species like us, there is survival value in caring, and displaying it has become part of the animal ritual called 'lekking'. These are the roots of the vast collective effort we put into our health services, yet they allow long-term genetic defects to persist. Birth control has been unsuccessful in slowing population-increase during the last 100 years. Raising the standard of living, which accompanies a falling birth rate, by building the equivalent of a new London every five weeks has environmental consequences that force us to consider other ways of reducing our population.

## Chapter 13: Taking Human Life & Species, Aims & Immortality

We have to face the reality that before long we will have to kill people. It is already legal under conditions of war or enforcing internal law and order. Our biggest obstacle is that we all belong to one genetic species. We look at different kinds of species and consider the evident fact that we are evolving from a genetic species into a behavioural one. Then we ask ourselves whether reclassifying individuals would help us come to terms with such an obstacle.

We all share a common gene pool and the Earth; we need a common aim that is acceptable to all; I suggest that this should be conserving information, since information is our immortality.

## Chapter 14: Law & Politics, Reputations & Punishments

All of the questions we have considered can be reduced to just one: 'Do we, or do we not, want to see the inevitable collapse of the human population to be orderly under the rule of law?' If the answer is yes, then we must persuade our politicians to pass laws that will achieve this. We consider the different options for passing rational measures into law, and also what these measured could be.

## Epilogue: In the Long Run

Flights of imagination into the past and the future give us hope – so long as we retain doubt.

# Chapter 1

## THE PROBLEM

Every two seconds, five more babies are born than people die; that is about 80 million more people to feed, house and employ every year. And there are already more than 7.3 billion of us.[1]

The words 'millions' and 'billions' are distinguished by just two letters. When spoken, the difference between them is a puff of air between the lips. No wonder newsreaders occasionally muddle them.

How long, in larger units of time, is a million seconds? How long is a billion seconds? If a line of people walked past you at one per second, a million people would take 11 days to pass. A billion people would take 32 years. You might find this image useful when next you meet a billionaire.

While on the subject of time, a tsunami at sea arrives in hours and is over in just minutes. Rescue takes days; clearing up and rebuilding take a few years. The population tsunami that is bearing down upon us has taken centuries to develop, and is taking decades to break. There will be no rescue because everyone will be affected. The only consolation is that clearing up will be done quite quickly – by the mechanisms of evolution and ecology.

Overpopulation is a real threat that is compounded by expectation. Article 25.1 of the United Nations (UN) Universal Declaration of Human Rights states:

> 'Everyone has the right to a standard of living adequate for the health and well-being of himself and of his family, including food, clothing, housing

[1] The United Nations announced that, on 31 October 2011, the world population had passed 7 billion (http://www.worldometers.info/world-population/).

> and medical care and necessary social services, and the right to security in the event of unemployment, sickness, disability, widowhood, old age or other lack of livelihood in circumstances beyond his control.'[1]

This was declared in 1948, when there were less than 2.5 billion people on Earth, and in most of the *developed* world few families had a car,[2] refrigerator, washing machine or central heating, or ate much exotic food, though most staples were imported. Today (2015), we are 7.3 billion, and the majority of us in the developed world regard a car, refrigerator, washing machine, dishwasher, television, personal computer, mobile telephone, central heating and fresh fruit and vegetables throughout the year as at least highly desirable, if not essential; add to this state-provided healthcare, education and security. The majority of people living in poorer countries can only dream of this level of luxury, and they do.

What do we, today, really mean by 'a standard of living adequate for health and well-being'? For the purposes of illustration only and to avoid arguing about it here, let us take it as being the average for London. Though there are many who enjoy a far higher standard, there are more who do not have the basic services and equipment listed earlier.

Think of London as a product of human endeavour. Think of it as a feat of civil engineering and social construction. Think of its monuments and public buildings, houses and offices, bridges and tunnels, hospitals and schools, roads and buses, lorries and cars, railways and trains, waterworks and plumbing, drains and sewage plants, street lights and wiring, refrigerators, television sets, radios, computers, beds, linen, kitchens, tables, chairs, cutlery. Think of training the teachers and doctors, solicitors and police officers, engineers and train drivers, firefighters and administrators who make this mighty city run smoothly.

Think of the countryside that was once cleared of wildlife to grow the food the people of London need, and which must be brought in daily; think also of the waste that must be taken out, and the nearby countryside that must absorb it. Think of the power stations and energy the city consumes, and the distant parts of the world that must be plundered to supply it. Think of the industries by which London earns its living, and the raw materials that must be produced to feed those industries.

Now find a piece of unspoiled countryside with a tolerable climate somewhere in the world, and think of building and equipping a new London on it. The cost and difficulty, and the materials and labour required are unimaginable, let alone training more than 25,000 police officers, 165,000 National Health Service employees, and staff for 3,000 schools, colleges and universities.

But if our concept of an adequate standard of living is reasonable, and if Article 25.1 of the UN Declaration has any meaning, it is necessary for the world to build the equivalent of a new, fully equipped and fully staffed London *every five weeks*, *before* we can start helping those of us alive today who do not have an overall adequate standard of living. Eighty million more people are born each year than die; this is 8 million people every five weeks – the same as the population of London.[2]

Eventually, as those among us who live in poverty die off, we who are left would need to build a new London every few months then at longer intervals. But this would only be effective if our rate of increase had begun to decline. By the way, about a billion girls alive today are under 15 years old, and all but a few will demand their right to have children. And still, people say that population will fall when we all have an adequate standard of living.

Madness.

Of course, we would not build each new London as a single unit, but as extensions to existing cities; nevertheless, the image is relevant and useful here. Maybe my assumptions are wrong

[2] 'Developing countries will be building the equivalent of a city of a million people every five days from now until 2050' (Royal Society, 2012: p. 7). This report does not say whether this is to accommodate the expanding population, or to also improve the conditions of those already in poverty.

and we do not need to build to the average standard of London, only that of a shanty town in a developing country. Really? Would you, who are reading this book, be content to live in one? Or maybe we only need to build a new London every two months, or three. OK, let us change the figures but it still does not alter the argument.

A weakness in any prediction is that the future is unknowable and that, like the weather or the stock exchange, the further ahead one forecasts, the greater the margin for error. Nowhere is this truer than when seeking to do good,[3] which is to say that trying to help people a generation ahead is a high-risk activity – not least because their perception of 'good' may not be the same as ours.

Frequency is not a safe predictor of probability. In other words, just because something has happened in the past, there is no *certainty* that it will happen again in the future.[3] The fact that, in the past, the Earth has always provided us with our food and raw materials and absorbed our waste does not mean it will always continue to do so. It is more sensible to use the frequency with which particular events have happened in the past as a *guide* in assessing the *likelihood* that they will occur in the future under defined conditions. This is how insurance works. I know of no flourishing insurance company or firm of actuaries that relies on gut feelings or 'God tells me' when assessing probability.

Predicting what the human population will be a generation hence cannot be exact. All we can do is to estimate what it was at various points in the past and discuss probabilities of it being something else in the future. Here are reasonable estimates of our population over the last half-millennium; they are shown as sets of two figures, the first figure representing the years of the Current Era (CE) and the second the millions of people: 1500: 458, 1600: 580, 1700: 791, 1800: 978, 1850: 1,262, 1900: 1,650, 1950: 2,521, 1999: 5,978, 2008: 6,707,[4] 2011: 7,000, June 2015:

---

[3] Believing that it can is the flawed base on which Matt Ridley founds his argument in *The Rational Optimist: How Prosperity Evolves* (2010).

7,320. The figures are more interesting when drawn as a graph (Figure 1.1).

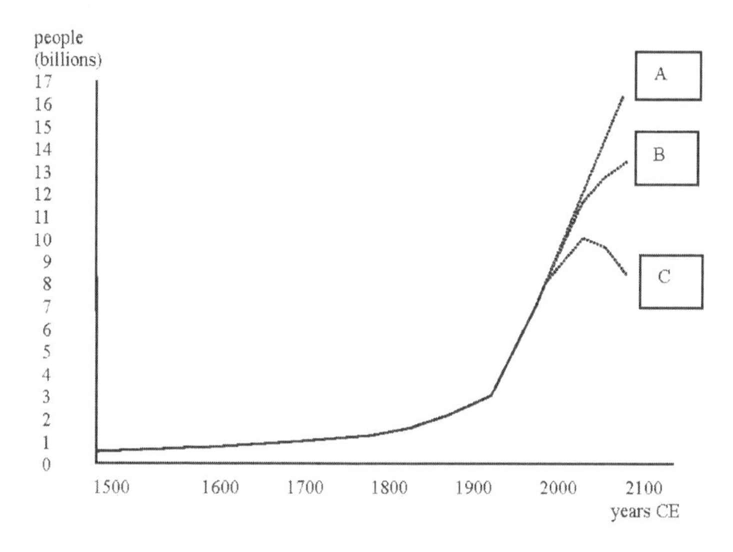


**Figure 1.1** Estimates of our population over the last half-millennium.

The lines in Figure 1.1 are based on two classes of information: the solid line follows the figures given earlier; they come from reasonably accurate censuses for some countries in modern times, and approximate estimations for other countries and in previous ages. We can take the figures as being tolerably reliable representations of reality.

The broken lines are quite diffcrent; they are predictions of three out of many possible trajectories, and they are based not on evidence, but on different probabilities:

A. The UN High Variant shows the population reaching 10.868 billion in 2050, and climbing to 13.556 in 2075, and then 16.641 in 2100. This is a nightmare scenario because it appears to disregard the principles of evolution, ecology and human behaviour. Mercifully, it is unlikely that we will get that far because we will run out of resources, or accumulate too much waste.[5]

B. The UN Medium Variant predicts 9.551 billion in 2050, and 10.409 billion and 10.854 billion in 2100. The curve shows the rate of increase slowing down but not declining.
C. The UN Low Variant has the population peaking at 8.342 billion in 2050, falling to 8.016 billion in 2075 and to 6.750 billion by 2100. We can achieve the orderly retrenchment indicated by this curve. However, I fear that the timescale does not pay sufficient attention to the principles listed in point A, and that is my reason for writing this book.

In the past, population alarmists like Thomas Malthus and Paul Ehrlich were wrong in predicting dates.[6] Malthus, writing just before the Irish potato famine, neglected migration to North and South America, Australia and Africa; Ehrlich failed to predict the Green Revolution in crop breeding, oil-based fuel, fertilisers and pesticides that vastly expanded food production.

In the 21st Century, there are no more lands to colonise without imperilling ecological services. Expanding agriculture depends on genetic diversity in crop plants, availability of fertilisers, fuel oil and new pesticides to which pests are not immune and which do not poison us. If these essentials are not presently nearing exhaustion, they inevitably will be – soon.

I remind you of the bored shepherd boy who amused himself by running back to the village crying *'Wolf! Wolf!'*, even though he had not seen one. Ever since Malthus, people have cried *'Wolf!'* over human population. However, I beg you remember the end of the shepherd boy story: the wolf came.

**********

I do not think our population will reach 16 billion people. It seems more likely that the ecosystem of the biosphere will stop supporting our excessive demands upon it before then.[4] Much

[4] In 2009, Columbia University Professor Jeffrey Sachs, special adviser to UN Secretary General Ban Ki-moon on the Millennium Development Goals, called on world governments to reduce population growth and work together

more likely is that reduced living space and food will induce us to change our behaviour. Whatever that behaviour may be, it will involve the question, ‘Do I compete or do I cooperate?’

One thing is certain, and it is that our population will eventually fall to maybe as few as 1 billion people with a standard of living about the same as that of today’s developed countries.[7] Meanwhile, at least 1 billion of us will die anyway during the next human generation of 35 years. Either most of us can die full of years and under the rule of law, or we can die messily, many while still young, as governments go rogue or collapse and warlords scramble for diminishing food and space. Where reality comes to rest between these two extreme scenarios depends on the laws we persuade our governments to pass soon. Laws to limit population will necessarily be tough but the alternative is worse.

The migration of about half a million people into Europe during the summer of 2015 is not the only fore-ripple of the population tsunami. There is also the race for finite resources: cash- and oil-rich states are buying up farmland and the rights to exploit freshwater, oil and minerals wherever they are for sale. Having lost between 18 and 45 million people through starvation and the associated violence during Chairman Mao’s *Great Leap Forward* (1958–1961),[8] Chinese people clearly appreciate the coming threat and have made at least some attempt at population limitation. The USA has been waging war in Asia ostensibly in its war on terror, whatever that may mean, but in fact to protect its supply of oil. Mayhem throughout much of Muslim lands is more complicated to explain, though its deeper causes may well have roots in the strife being on lands where Eurasian farming and the three main monotheistic religions began, as we will discuss in Chapters 6 and 7.

States with dictatorial oligarchies that keep the bulk of their people in poverty are acquiring, or have recently acquired,

---

to keep climate change from causing an immense human catastrophe, starkly warning that: ‘We’re on a trajectory that is absolutely unsustainable and profoundly dangerous.’ http://www.un.org/apps/news/story.asp?NewsID=32074.

nuclear weapons (Iran, North Korea, Pakistan) and, whatever the world may wish, only the politically naive believe that such dictators will not either use these weapons, or make them available to international terrorists, when mass starvation is pandemic. I think that, unless rich people share their wealth with the poor, and the prolific reduce their rate of reproduction, a major city in an industrialised country will be destroyed by a nuclear weapon before 2050. I hope I am wrong, but remind you that thinking and hoping are unrelated.

It is useless to pray for some benign dictator or epidemic disease that will take away from us individually the unpleasant responsibility of making heartbreaking decisions of life and death. A benign dictator is too rare to be practical, and epidemic diseases are largely indiscriminate – at least with respect to human behaviour.

I shall argue that we, as a species, are a natural member of the Earth's fauna and, like all other living things, subject to the mechanism of evolutionary selection, which is never indiscriminate. We, and all that we do, are also ultimately controlled by ecological mechanisms; that is why we need to understand how both work.

Not only is the crisis facing us real, it is also urgent. As we fly over even the crowded islands that are Britain, we can still see lots of open countryside. Indeed, this is true in much of the developed world, especially the USA. With so much obviously undeveloped land between towns, few people regard the threat of overpopulation as serious. It is seldom in the front of our minds that each one of us in our houses needs a lot of natural countryside for environmental services. These include supplying us with clean air and water, and absorbing our sewage, rubbish and pollution, let alone providing us with places where we can holiday.

Here is a parable. We are like unto duckweed.[5] Duckweed is a small green plant consisting of a flat leaf about 3 mm in

[5] The duckweed analogy has been adapted from Wilson E. (1998: p. 320). If each duckweed leaf is 5 $mm^2$ in area, multiplying it by $2^{30}$ gives about 5,000

diameter, with a tuft of fine roots beneath, which floats, often in large quantities, on still water. So long as there is clear water, the duckweed goes on growing and multiplying. Only when the available surface is completely covered does murderous competition between the little plants begin. Certainly, there is plenty of room when the water is only half covered. Now imagine a square lake with each side measuring 100 m. The area will be 10,000 square metres ($m^2$), or one hectare. Now we introduce one duckweed leaf onto it.

If each leaf grows to full size and divides in one day, the lake will be half covered in 30 days – that is, there will still be 5,000 $m^2$ of clear water. On which day will the lake be fully covered and individuals begin to die in rotting heaps?

On the thirty-first day.

Our human population is well into the equivalent of its thirty-first day.

Whatever the figures, we need to answer four key questions:

- Should we try to build, equip and staff the equivalent of a new London every five weeks?
- What are the human consequences if we do not?
- What are the environmental consequences if we do?
- How will either of these consequences affect future generations?

---

$m^2$. Bodanis (2000: p. 166), gives a similar 'pond in your garden, with a lily plant floating in it that doubles in size every day. In eighty days the lily entirely covers the pond'. Lily leaves are approximately 400 $cm^2$, so $400 \times 2^{80}$ $cm^2$ suggests an improbably large garden.

# Chapter 2

## THINKING ABOUT ANSWERS

Chapter 1 ends with four enormously important questions, so it is essential that we call on all our shared mental skills in trying to answer them.

There are many ways of thinking about thinking, and each of us will prefer the one that suits them. A system that appeals to me divides all questions into three categories, according to how they are answered. Questions in one category invite answers that refer to feelings; others in a second category ask for rules, habits or customs; and a third refer to the evidence of our senses and rational thought.[6] Of course some questions need to be answered by referring to two or all three of these categories, but they can be broken down into simpler questions, each fitting one of the basic three.

Of course there are other ways of classifying questions. The four questions at the end of Chapter 1 are of two distinctly different kinds. One kind, 'Should we try to build the equivalent of a new London every five weeks?' is a moral question – it calls for a decision. The other three questions seek predictions. Forgive me if the next section appears pedantic but, to develop my arguments, I need to be sure that we agree on the main areas of understanding. By all means skim-read or skip the next section if what I offer appears self-evident.

---

[6] These three categories appear in several authors, for example, Sober & Wilson E. (1998: pp. 221–223) and Francis Bacon's 'idols' in Gould (2000; pp. 54–55). See also Miller (1986: pp. 103 and 125) and Wilson E. (1998: pp. 34, 116 and 228), although on p. 125 he offers a different classification, as does Gordon (1990).

## 1: Questions that can be answered by referring to feelings

We share basic biological feelings and emotions with most mammals: hunger and repletion, alertness and tiredness, curiosity and sexual drive, fear and a desire to excrete. Unlike them (as far as we know), we can live also for the joy of more cerebral emotions, such as those elicited by Cezanne or Sibelius, understanding a mathematical theorem for the first time, enjoying exquisite food, appreciating a sunrise over a winter forest and revelling in a brilliant sporting achievement. These are what some call 'higher feelings',[9] to distinguish them from what we imagine animals may feel. On a different, though not lower, plane are the feelings we get when we offer service, cooperation and trust; and there is increasing evidence that we also share many of these feelings with animals. Most of us also seek to reduce feelings that involve fear, suspicion and outrage, hatred and distrust, dishonesty and deviousness.

The communications media constantly remind us that feelings are a better way of attracting attention than imparting factual information or arguing reasonably. We are appalled by pictures on television of people less fortunate than us – spindly limbs and bloated bellies amid dust and skeletal animals. But the media seldom discuss *why* these people are in such distress.

That powerful but slippery word 'why' can be answered by referring to any one of feelings, rules and logic, but only the last can lead on to deeper levels of understanding. Thinking rationally about it reveals that the word 'why' has several levels of meaning. The obvious reason why babies starve is because they do not have enough food. Why don't they have enough food? Because, for example, the rains failed. Why did the rains fail? Because there were no forests to cool the land and so allow rain-bearing clouds to reach the region. Why were there no forests? Because people cut down the trees for fuelwood and animal fodder. Why had they consumed so much wood and fodder? Because aid and modern medicines caused many of their children to survive where previously they would have died. In other words, aid from developed countries allowed the people to simply breed beyond the capacity of their land to supply their demands.[7]

Such logical enquiry offends our feelings, and we regard it as tasteless, if not offensive. We feel that it is more important to alleviate the pain inside us as quickly as possible,[8] and the easiest way to do that is to give money.[9] We disregard the chilling evidence that these impoverished populations are increasing too rapidly for sustainable relief, and that aid merely defers the problem, greatly enlarged, to future generations.

If this is the case, why do we have such sympathetic feelings? If we again peel back the layers of 'why', the answer is both complicated and understandable. Let us follow what actually happens when we see an image on television of children dying in a drought. The first thing we feel is 'feelings' – it does not matter whether they are of outrage or sympathy, they are feelings. And they are caused by nerve impulses from our eyes and ears travelling to our brains and other impulses travelling from our brains to special organs that release certain chemicals into our blood, and these in turn cause us to feel as we do.

The next level of 'why' is why are these particular feelings disagreeable? And the answer to this is because the nerve impulses that pass during this particular sequence interact with two sources of memory (prejudice, if you like): memory of our previous experiences – including how others have reacted in our presence to similar scenes – and the behaviour that is controlled by our genes. In contrast, a different set of impulses from the eyes and ears, say of an obviously affectionate couple in a

---

[7] In fact, this is a simplification. As I shall explain in later chapters, the causes of our runaway success as a species stretch back far into the past, and have their roots in our unparalleled capacity for cooperating with one another.

[8] Wilson E. (1998: p. 325): 'No one can seriously question that a better quality of life for everyone is the unimpeachable goal of humanity'; see also Cohen (1995: p. 292). However, Ridley (1996: pp. 21 and 137): 'The more you truly feel for people in distress, the more selfish you are in alleviating that distress'.

[9] See Buarque (1993: p. 49) for a savage riposte.

tranquil setting, would elicit quite different feelings – probably of what we would call happiness.

The third level of 'why' asks why do our genes cause us to feel outrage rather than happiness? The answer to this is that the appropriate feeling causes us to act in a way that eventually results in either us surviving as an individual, or copies of our genes surviving in other living things.

The fourth and final level of 'why' asks why our genes should be programmed in this way. Because, when successive generations of our ancestors experienced feelings suited to the particular stimulus, they acted in a way that increased their chance of surviving; this allowed them to leave more offspring than individuals that did not survive as long. Because genes influenced at least some of their actions, these genes accumulated in successive generations and we inherited them.

To recap, we can label these four main levels of answer as follows:[10]

A. *Reception*: nerve impulses from sense organs to an effector organ via the brain and under the moderating influence of memory and genes.
B. *Action*: the release of chemicals; or, when movement is involved, muscle contraction.
C. *Survival value*: the action we take either ensures that we as individuals survive together with our genes, or that action enables another person to survive, together with copies of the genes they share with us;
D. *Evolution*: our genes that survived because of the above steps were inherited by our descendants.

Genes and the feelings they produce cannot predict but thoughts can.[10] Should we not question the usefulness of our feelings as we try to answer the London questions?

[10] Dawkins (1989: p. 201): 'We, alone on earth, can rebel against the tyranny of the selfish replicators.'

## 2: Questions that can be answered by referring to rules

The first 'why' can be broken down into two parts: the moderating influence of (1) memory and (2) genes. Memory in this case is of two kinds: (1) personal first-hand and (2) social, that is the rules of the society we grew up in.[11] Clearly the first is profoundly influenced by the second. Such rules take many forms: laws we must obey, some more serious than others and attracting greater penalties for infringement; government 'guidelines'; customs; prejudices; habits; manners; even what we call common sense.[11] It is a wide spectrum, and it ranges from those that are essential to the running of society to those that have persisted with us into adulthood because we seized upon them uncritically when we were children. Many of these are the rituals our parents performed, and our unconscious childish reason argued thus: since our parents have been so good at the business of life that they have succeeded in rearing us, a safe bet in this confusing world is for us to do as they did. Anyway, it does not matter how we tie our shoelaces, and it saves us working it out anew each morning.

Sometimes, we are brought up to have absolute faith in the rules our parents taught us, and we would feel like fighting anyone who disparaged them. Other more liberal parents encourage their children to ask questions and so set them free to observe the world as they (the children) perceive it to be, and to make their own decisions. An important part of this freedom is the rejection of absolutism because it obstructs all ways of thinking.

I was deeply alarmed by an Englishman living in Spain in 1994 who began his career as a woodwork teacher, moved to remedial English for junior classes, rose up through the ranks of a teachers' union, and was seconded onto various government committees to advise on legislation. He said to me: 'I have sat through days and days of evidence and didn't care *that*', he

[11] '... that overrated capacity composed of the set of prejudices we acquired by the age of eighteen ...' Wilson E. (1984: p. 120).

snapped his fingers, 'for one word of it; I know that what I feel is right, and advised the Minister of Education accordingly'. Lucky him. Clearly he was seldom burdened with logical thought.

In November 2000, my wife and I were on a flight from South America. Our seats were by a window near the front of a section of a passenger compartment; beside us was a young Lebanese Muslim, he told us later. Across the aisle and in the front row of seats was a Sephardic rabbi in a frock coat, with a younger acolyte and another hatted Jew. They carried out their rituals with much ostentation – standing up from time to time, reading from a book, washing and bowing. The young Muslim turned to us and hissed, 'Look at them! Look at them praying! And they shoot children!' When the Jews had finished the obvious parts of their ceremonies, the Muslim laid his prayer mat to the side and a little in front of them, faced what he took to be the direction of Mecca, and began his prayer rituals. I thought at the time that the only way these Jews and this Muslim could settle their differences is by fighting – both know that they are right, so there is nothing to discuss. Oh how we hate one another for the love of God.[12] I wonder how many more people are cursed with this certainty.

All three of these ways of thinking – carpenter, Jew and Muslim are controlled by rules each has been taught, and which each knows to be ultimate truths. Each is convinced that the salvation of the world can be achieved only in *their* way. Faith is distinct from rational thought in that it has no objective evidence to support it. It usually starts from what a person has been told or read, so is based on hearsay. There can be no discussion about an article of faith. 'I believe' is the end of a conversation. Faith is a powerful political force, made stronger by persuading the faithful that they cannot submit themselves to the debasing test of rationality.[13] Small wonder then that it is said that convictions are more dangerous enemies of truth than lies.[14]

People are often disconcerted when they ask me whether I believe in Darwin's theory of evolution by means of natural selection and I reply, 'No'. They are relieved when I go on to explain that, to me, the word *believe* is used to mean several different things: for example, 'I believe it will rain tomorrow', or 'I believe there is no God but Allah'. I *think* that Darwin's theory is the best explanation we have for what we observe in the natural world; but 'believe' – no. If somebody comes up with a better theory – 'better' in the sense that it stands the tests of observation, experiment and reasoning more successfully than Darwin's – I would happily accept it. So, incidentally, would have Darwin because he was deeply troubled by the effect his theory had on Emma, his profoundly religious wife, who feared that his ideas would prevent them from one day being together in Heaven. In fact, Darwin's theory has been subjected to a more sustained and critical assault on it by more people determined to show that it is flawed than any other theory in the history of science. They have all failed – so far.

I do have faith. I believe, with all the certainty in me, that there is an understandable connection between cause and effect; that there are no mysteries, only areas of ignorance and incomprehension. This is the tenet of a philosophy called *naturalism*. A different and much more universal tenet is called *humanism*, and it holds that every individual member of the human species has a unique and sacred nature, which is fundamentally different from the nature of all other animals and all other phenomena. Humanists believe that our collectively and individually unique nature is the most important thing in the world, and it determines the meaning of everything that happens in the universe. We will look at another philosophy in Chapter 8: *The Special Case of Islam*. The only way to resolve the differences between these and many other philosophies or articles of faith is to examine them logically.

The origins of these faiths light up the differences between them. As we have come to understand more and more about the universe, so the amount of detail we must take on trust has diminished. All animals perceive the world about them, and their reactions to it are strictly practical because, as far as we can tell, they seek no more than to survive individually and to breed. One of several features distinguishing us human beings from other animals is that we try to explain what we perceive.

When we fail to do this satisfactorily, we construct myths to explain our perceptions. The capacity to imagine these myths seems to have evolved after we had begun to speak but before recorded history. It probably happened rather quickly, and has been called the *cognitive revolution*.[15]

A possible sequence of cause and effect that led to the evolution of imaginary explanations runs as follows: family groups of our primate ancestors fought each other for access to environmental resources and mates. The chance of winning such battles was increased if greater numbers of male fighters cooperated together.[16] Language evolved to enhance social bonding in larger groups, but the greatest number of people an unaided human voice can address is about 2,000.[17] If listeners could be persuaded to talk among themselves about what they had heard the speaker saying, the message would spread. If this message related to something already within the listeners, such as a desire to understand a phenomenon or achieve a collective or individual aim, they would feel a sense of unity with the messenger. Such a message can grow by accreted input from those who pass it on until it assumes the form of a narrative.[18] The narratives of Norse or Ancient Greek legends and deities took the form of a normal dysfunctional human family or soap opera.[19] Other narratives describe how the people that believe them came into the world, and yet others prescribed a moral code with or without practical rituals.[20] Religious rituals often arose out of eminently sensible hygienic precautions such as washing.

The survival value of belief in such myths was that they bound people together to form armies, states and civilisations, and each level of these was better fitted to defending its possessions and assumptions from smaller entities. Examples of these myths include religions, nationalism, socialism, capitalism, communism, conservatism, consumerism, scientism, naturalism, humanism and many other sets of ideas often characterised by the suffix '-ism'. These titles are no more than attempts to classify the many different ways in which people try to make sense of the impressions they received about the world outside and inside themselves. Even though each '-ism' excludes parts, if not all, of the others, they are seldom mutually exclusive. All of them are myths in the sense that they contain elements that do not satisfy the philosophical test of proof. All myths require an act of faith by followers. No one can prove that there is or is not a god; even naturalism cannot prove that every phenomenon has a cause, let alone that it can be understood by the human mind. Of course, one cannot prove anything, least of all an absence; one can do no more than *disprove* claims and assertions.[21]

## 3: Questions that can be answered by referring to the evidence of our senses and logical argument

This category has three main difficulties: (1) the margins of answers are blurred – 'Is it raining?' 'No, drizzling'; (2) what we perceive is often modified by mental models already within us; and (3) 'factual' answers change.

The first is straightforward, but the other two need explanation. Mental models can be strikingly illustrated by a simple experiment. Figure 2.1 shows two images. What do you perceive when you look at them?

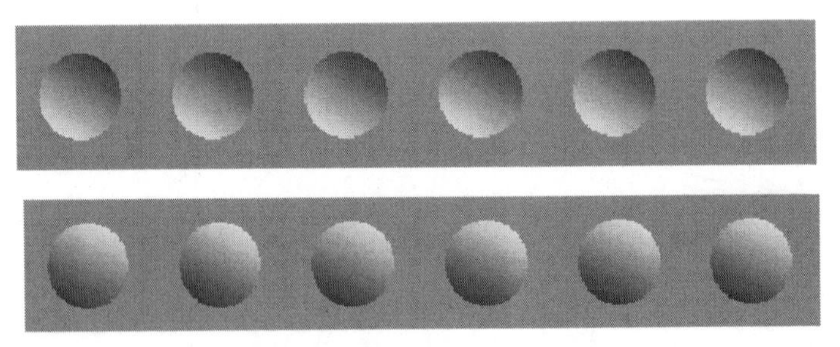

**Figure 2.1** The mind adds information that is not in the image.

Most people describe the upper picture as a row of cavities, and the lower one as a rather similar row of protuberances.[22] Agreed?

Now rotate this book so that it is upside down.[12] What do you perceive now? Most people say that what was the lower picture and is now the upper is no longer a row of protuberances but a row of cavities; and what was the upper picture and is now the lower one is no longer a row of cavities but a row of protuberances.

Notice that I do not ask what you *see*, only what you *perceive*. In fact, the two sets of information your eyes sent to your brain, one before and the other after rotation, are identical. Your mind has *added* the information that light usually comes from above and that the lower surfaces of cavities are illuminated while the upper surfaces are in shadow, and the reverse is true of protuberances. In preparing for what you *expected* to see, your mind made a model, and that was why you *perceived* a distinction that was not there in reality.[13]

You may have seen the following experiment (Figure 2.2), in which case your software (shorthand for the way the mind works) has been prepared. If you have not, read it quickly:[23]

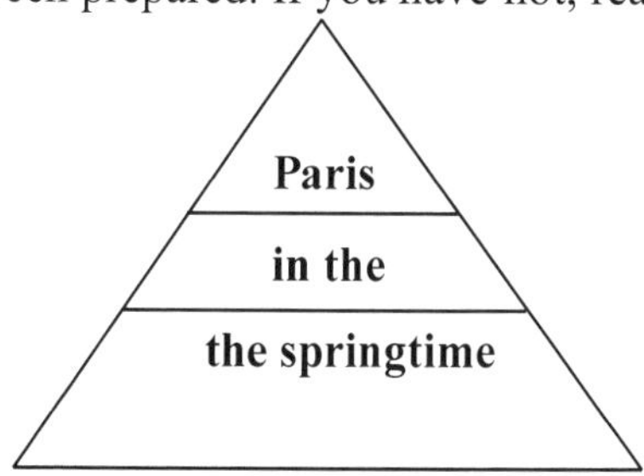


**Figure 2.2** The mind removes information that is in the image.

Did your mental model about the rules of syntax remove the second 'the'? This shows why proofreading is so difficult to do well, and why your computer often warns you about repeated words.

[12] If you are reading this book on a handheld tablet or laptop computer, you may find the image rotates as you turn it; if it does, try shining a light across the screen from the lower edge of the frame.

[13] Pinker (1997: p. 28) and see the index for 'Modularity of mind'.

These simple tests show that we sometimes find it easier to reject the information our sense organs give us than to change our preconceived models to fit the reality of what our senses are telling us. If the human mind can add to or remove from the simple information it receives as in these examples, how much more can it distort complex social input, especially after a lifetime of indoctrination by the society in which one grew up? Another important point is that, however hard we may find changing our minds, we each acquired these models after birth. Even though they can become hardwired into our neural pathways, they can be changed with rational training.

**********

I came across the third difficulty – how what we think of as the truth can change – when teaching elementary chemistry in schools. I described the discovery of *relative atomic masses* (RAMs), as follows, though with suitable preamble, demonstrations and practical exercises:

Oxygen and hydrogen are elements but water is a compound. The smallest identical particle of an element is an atom, and the smallest identical particle of a compound is a molecule.

When you put two wires into water and pass an electric current between them, hydrogen bubbles off one wire and oxygen bubbles off the other. The volume of hydrogen given off is twice that of oxygen. With rather more sophisticated apparatus than we have here, you can calculate that equal volumes of gases at the same temperature and pressure contain equal numbers of atoms (or molecules, if the gas was a compound). Therefore, from the electrolysis of water, you can reason that two atoms of hydrogen had combined with one atom of oxygen to form water or $H_20$.

Now you can weigh your volume of oxygen and compare its weight with that of half (to make the same number of atoms) the hydrogen you collected, and you will find that 1 atom of oxygen is 16 times heavier than 1 atom of hydrogen. Weighing something is a way of estimating its mass. So this experiment indicates that the mass of an oxygen atom is relatively 16 times more than that of a hydrogen atom; that is, its RAM is 16. By a range of other methods, scientists have measured the RAMs of all other elements and compared them with each other and with hydrogen. They found that RAMs *were whole numbers*. This was a huge step forward in understanding because, in chemistry and everyday physics, atoms cannot be split up – they react with each other as whole entities. This gave a sound theoretical basis for most of what is taught in school chemistry today.

Scientists then measured RAMs more carefully and found that some *were not whole numbers*; for example, the real RAM of chlorine is about 35.5. Which is to say (I remind you) that the mass of an atom of chlorine appears to be about 35.5 times greater than that of a hydrogen atom. This made nonsense of the idea that atoms react together as whole entities. Then scientists found that, in fact, chlorine atoms occur as a mixture of two forms, one of which has a RAM of 35 and the other of 37, and that these two forms occur in such a proportion that the average RAM of chlorine appears to be 35.5. Thank goodness for that, because the whole atom theory was saved, and now RAMs *were whole numbers* again. $^{35}Cl$, which is the way we write chlorine with a RAM of 35, and $^{37}Cl$ are called isotopes of chlorine, and we cannot readily separate or tell them apart chemically. Isotopes of some elements are radioactive, and study of them opened the door to our understanding of nuclear physics.

Then scientists measured the RAMs of radioisotopes even more carefully with modern methods and discovered that their RAMs *were not exactly whole numbers*, and this led to our understanding of quantum physics and of the deeper nature of the universe. I would not be surprised if that is also not 'wrong', but it will do as an answer to be getting on with in our present state of knowledge.

Truly, education is a progression of diminishing deceptions.

So, I beg you beware of people who talk about 'facts' and 'truth'. It is more sensible to treat all knowledge as approximate and provisional. There *is* an absolute, and it is the information that determines how an entity is constructed and how it behaves. When we understand some of that information and convert it into human symbols, we call it knowledge.

The television world is obsessed with correct answers. Many quiz programmes stress the glory of getting it right – or at least what the setter of the question thought was right, and above all quickly. The glib politician's sound bite is seen as correct, if only politically. It does not bear thinking of the millions of institutionalised children (well, they are at school) that are labelled as unintelligent because they did not give the 'right' answer. I am impressed by the physics candidate who, when asked how he would measure the height of a building using a barometer, said that he would drop it off the roof. He would then calculate the height of the building in metres by squaring the time in seconds taken for the barometer to hit the pavement, multiplying the result by 9.8 and dividing that by two.[14]

**********

Putting ways of thinking into three baskets, feelings, rules and rational thoughts, fits well with another way of looking at our minds. Eric Berne (1910–1970) worked as a psychiatrist in California and asked himself what actually transacted between people when they spoke to each other. He called his method *transactional analysis*.[24] He noticed that people appeared to behave as if they were always in one of three *ego states*. Berne called the three ego states *Parent*, *Child* and *Adult*, and he noticed that they could flip from one to another instantly. When a person was in the Parent ego state, they felt, thought, talked, acted and responded just as one of their parents had treated them when they had been between the age of two and five years. In Berne's time, most people lived with two parents when they were growing up, so an individual's Parent ego state was divided in two, and they could switch between them. The Parent ego state is characterised by telling others what to do. When the person was in the Child ego state, they were inquisitive, sparkling, creative, naughty, whining, grizzling and laughing. When a person was in Berne's Adult ego state, they objectively appraised what they sensed and calculated the effects of

[14] Between 1972 and 1978, I was a senior examiner for the University of Oxford Delegacy of Local Examinations, Ordinary Level (15–16 year olds), with an entry of about 20,000 candidates.

consequent actions in the light of their previous experience – they functioned more like a computer than a feeling human being.

When two people interact they can do so Adult-to-Adult (A–A) and this gets things done. They can also interact Adult-to-Parent (A–P), Adult-to-Child (A–C), P–A, P–P, P–C, and so on, giving nine possibilities. Sometimes an opening approach could be A–A '*Where are my cufflinks?*' But the response, *'Why do you always blame me for everything?'* is C–P. It crosses the approach and results in slanging or sulks. Or the response could be *'You are always losing them'*, which is P–C – again a crossed communication with the same results. A further complication is that interactions need not be honest. '*Please Aunty, can elephants swim?*' is seemingly an A–A transaction, but the nephew is really trying to catch his aunt out, so is really C-P. Berne calculated that there are 6,480 types of duplex transaction; fascinatingly, only 15 of them commonly occur in ordinary transactions, the rest are largely academic. I commend deeper reading of this subject, and assessing what part such an approach to thinking might have in answering the London questions.[25]

How are we going to apportion these three influences when we consider the London questions? There are many people who will always take the view that the only way to approach moral questions, like whether we should build a new London, is to rely on gut feelings, which is what they instinctively feel is 'right'. Others are more cautious and draw up a balance sheet of evidence and argument for and against each aspect of their final answer, which is what I am trying to do in this book; and some depend on precedent and rules. All three ways of thinking have the same political weight in a one-person, one-vote democracy. Is this the best way of planning for the future? It is certainly better than all others we have tried.[26] But we have not tried them all.

A consequence of muddling rules, rationality and feelings is that our ideas can seem to be attached to our personalities. Sometimes we make the attachment ourselves, and take criticism of the idea as an attack on us as individuals. At other times we ourselves are quite clear about the distinction, but those who disagree with us make the argument personal. This is particularly pertinent when biologists try to explain the roots of racism, and are accused of condoning it.[27] Equally, the suggestion that a human life should be ended deliberately can stand apart from the person who suggests it.

A good scientist, entrepreneur, administrator or industrialist continually searches for correction – '*I move towards the man who contradicts me: he is instructing me*', as the French Renaissance statesman and essayist, Michel de Montaigne, wrote.[28] It is the first duty of a scientist – and, indeed, everyone else who deals with reality – to find someone to prove them wrong.

**********

An effective way of helping ourselves to think rationally is by examining the evolutionary origins of our feelings, social rules and other behaviours. This is why I go into such detail about them. For example, we may be privately amused when we notice ambitious parents wangling an invitation for their child to play with another whose parents they regard as higher in the social scale. They would resent being told that monkey parents do it too. It is more parsimonious to think that, like much of our anatomy and physiology, this behaviour is derived from an ancestor we shared with monkeys 20 million years ago (mya). What other mental habits are that ancient? Are they suited to answering the London questions? We need to know.

Before going further, we also need to agree on the use of two words and the ideas that flow from them: genetic codes *evolve* and individuals *develop* from a fertilised egg. Our individual genetic code is the result of evolution and is fixed at conception. It changes little during our lives, whatever environmental conditions we encounter.[15] On the other hand, how we develop physically is strongly influenced by what we eat and experience. In a parallel way, our innate behaviour, which is genetically

[15] When it does change, the changes cause cancers and senility.

determined, remains relatively fixed throughout our lives. Quite different from this is our learned behaviour, which changes all the time. Remember that many of our mental models are innate.

The models in our minds are influenced, if not derived from two sources: our genetic inheritance and what we have learned individually during our lives. Even this distinction is blurred because much of what we learn when young becomes 'hardwired' – like what we inherit genetically.[29] Modern synthesis recognises additional layers of control in the way genes act. The combined effect is called *epigenetics*, sometimes referred to as 'evo–devo'.

There is another way of looking at our mental processes. When we think about pure mathematics, we do so with our minds only; we do not need to observe the outside world or do experiments. We call this kind of thinking 'before experience'. On the other hand, thinking about biology in this way is unsuccessful – we have to look at living things before putting a biological argument together. This kind of thinking is called 'after experience'.[16] It helps to understand that these two ways of thinking are separate, even though we cheerfully mix them when we are developing a set of ideas.

Since we receive no knowledge that is uncoloured by our interests and perspectives,[30] is reality really limited to what we can conceive?[31] No, the properties of the world can only become available to us through our sense organs and instruments; but this does not mean that things beyond our perception, even with instruments, do not exist. A recent and major stage in the evolution of our minds was to survive as hunter-gatherers in a world that was not much altered by us; it did not evolve to live in crowded cities, or to understand itself.[32] Yet we have made huge progress in both areas.[33]

**********

Thus far we have thought about general aspects of perception and other influences on the way we take decisions. Now I want

---

[16] They are correctly labelled *a priori* and *a posteriori*, respectively.

to tackle three more ways in which our perceptions can get in the way of clear thinking. The first way has to do with how we understand certain words; the second is our perception of the environment; and the third is how we perceive ourselves.

To understand what is going on in the environment or in ourselves, we have to use words. Like us, words evolved. Not only that, neither they nor we are perfectly adapted to performing the tasks expected of us – we both just muddle along as best we can under the conditions in which we find ourselves. For example, we use the phrase 'This person is an achiever'. The word 'achievement' is loaded with human intention, as are 'competition' and 'cooperation'. But they need not be. I will discuss them in the next chapter, although to understand them, there is no escaping some basic chemistry.

What has chemistry to do with answering the London questions? The first and most important of those four questions demands a moral answer. I shall argue that no one can see what morality really is without understanding some elementary chemistry. If you are to understand chemistry, you need to start somewhere, and it makes sense to start at the beginning, and the beginning of chemistry is also the beginning of time, and that is as good a beginning as any other. In looking at the steps we took to arrive where we are, right from the beginning of time, it is at least simpler to think that there are no mysteries – only gaps in our understanding. And this is naturalism – my article of faith.

In any worthwhile discussion, the two sides have to agree on their ground rules. Talking is pointless if one side claims a special link with a god that the other cannot question. So, if you are a devout believer in a god, I ask you, at least temporarily, to suspend your belief while reading on. Even though there is no scientific evidence for the existence of a god, neither is there evidence that one does *not* exist, so you may be right.

# Chapter 3

## COSMOLOGY TO CELLS, COMPETITION & COOPERATION

Edwin Hubble (1889–1953), the US astronomer, made one of the greatest contributions to human thought. He discovered that the visible universe was expanding.

By 'running the film backwards', that is, imagining it contracting, there was obviously a moment when it could contract no longer. That point was about 13.7 billion years ago, and is taken as the birth of our universe.[17] Before our universe was born, it was preceded by a *singularity*. In a singularity the laws of physics, as we understand them today, do not appear to exist. The question 'What happened before the laws of physics began to operate?' is meaningless, because there was no time in which it could happen. A singularity is thought to have been an infinitely small, infinitely hot and infinitely dense point.

In a simple way, we can express everything that we encounter in our universe in the four fundamental dimensions of *mass*, *energy*, *space* and *time*. In fact mass can be turned into energy by a nuclear reaction, and time changes with velocity, which is length (a measure of space) divided by time. So it is simpler to think of *mass-energy* and *space-time*.

Mass-energy and space-time began when the singularity became finite by beginning to expand. As soon as there was time, things could happen, and they happened rather quickly to start with; so quickly that it created an explosion called the 'Big Bang'. The temperature of the universe dropped from the singularity's infinite heat to 100 billion degrees centigrade in about 100$^{th}$ of a second. Three minutes later the universe had cooled to 1 billion degrees and consisted of quark-gluon

---

[17] There may be others.

plasma.[34] Out of this plasma, two fundamental particles formed: positively charged *protons* and negatively charged *electrons*. Then, when two protons collided in the plasma with enough energy, one of them lost its positive charge and turned into a third kind of particle: a *neutron*. This is a simplification, for we now know that protons and neutrons are made up of smaller units, but it will serve our purposes here.

Eventually the universe cooled enough to allow electrons to buzz around a proton with the positive and negative charges holding them together. One proton and one electron together form a hydrogen atom. One neutron, one proton and one electron form a deuterium atom. When two deuterium atoms, each with one neutron and one proton, collide with enough energy, they combined to form a helium atom, which therefore consists of two protons, two neutrons (the *nucleus* of the atom) and two electrons buzzing around them. This fusion process releases unimaginable quantities of heat, and is going on continuously in the Sun. It is something we have been trying to achieve on Earth to solve our energy problems.

Energy released by nuclear fusion of hydrogen to form helium via deuterium inside a star causes outward pressure and tends to increase the star's diameter. For most of a star's life, the expansive force of heat just about balances the contractive force of gravity. However, this balance eventually breaks down, and so gives stars a lifespan; and we can see different stages of it in the visible universe. When a star has used up 10% of its hydrogen and turned it into helium, it then uses up the remaining 90% as quickly as it used the first 10%. This later stage releases energy much more quickly, and the expansive force of heat production overcomes the contractive force of gravity and inflates the body of the star. When nearly all of the hydrogen has been converted into helium, heat production declines and gravity overcomes the expansive forces. The body of the star then collapses inwards, and the force of gravity heats the core so much that deuterium and helium atoms fuse into carbon atoms (six each of protons, neutrons and electrons). Various combinations of protons and neutrons then fuse together to form the nuclei of some heavier elements. When temperatures allow them to do so, electrons join the nuclei to balance the charges.

The last act in the life of a star begins when the core of the star collapses catastrophically under its own gravity, releasing energy into the outer layers, which expand rapidly and cool to a red colour. At this stage the star is called a *red giant*. The outer layers then collapse and crush the core to plasma, which then superheats, even by cosmological standards. In a large star, this results in the most violent explosion known, after the Big Bang, and it is called a *supernova*. For a short time, the star becomes as bright as a galaxy. The explosion tears apart the plasma of protons, neutrons and electrons and allows them to reassemble in different combinations, each one becoming a different element. Ash from an explosion such as this drifted into space and was caught up in the gravitational field of another younger star, which is our Sun, and gave us the 90 or so elements we find on Earth. As far as we can tell, there are no other naturally occurring elements in the visible universe.

The ash circled into a disc, and slowly coalesced into the planets of our solar system. At first, gravity and radioactive decay heated the infant planets, melting them so that the heavier elements – mostly iron and nickel – sank into the cores, and lighter elements – like hydrogen, oxygen and aluminium – formed a skin on the surface as the planets cooled. Eight of the nine planets[18] of our solar system are named after the gods of classical mythology: Mercury, Venus, Earth, Mars, Jupiter, Saturn, Uranus, Neptune and Pluto. By the way, a handy mnemonic for remembering the order of all nine planets (including the Earth) is 'My Very Educated Mother Just Showed Us Nine Planets'.

**********

Now some much simplified elementary chemistry, which will enable us to follow the reasoning that leads us to understanding

[18] In 2006, the International Astronomical Union formally declared Pluto to be a dwarf planet rather than a planet proper.

what we really are and, thence, how we can answer the moral London questions (Should we build one …?) confidently.

Electrons are the means by which atoms link together to form molecules. We can think of these links as hooks. The hooks work in two ways, and one way is for an atom to lose or gain an electron. A hydrogen atom can lose its electron and doing so allows the proton in the hydrogen atom to show its positive charge; likewise the electron can show its negative charge, like this:

$$H \rightarrow H^{+} + e^{-}$$

Another kind of atom, such as a chlorine atom, can gain an electron like this:

$$Cl + e^{-} \rightarrow Cl^{-}$$

Whenever an atom loses or gains an electron it is said to *ionise*, and the charged body so formed is called an *ion*. Oppositely charged ions attract one another and they join together to form a molecule, as shown here:

$$H^{+} + Cl^{-} \rightarrow HCl$$

HCl is a compound called hydrogen chloride. Notice that the positive and negative charges cancel each other. As soon as HCl dissolves in water, its ions separate again and the solution is called hydrochloric acid. Whenever a molecule of a compound releases hydrogen ions in watery solution, it is described as an 'acid'.

The second way in which atoms can hook together with their electrons is by sharing them. In its free state, hydrogen atoms usually share electrons in pairs, like this:

$$H + H \rightarrow H_2$$

We can imagine the two electrons orbiting around the two protons and holding them together. For our purposes, there really is not much more to chemistry than that: the rest, as they say, is vocabulary.

Within our solar system, only the Earth is at the right distance from the Sun for water to remain in liquid form, and large enough to hold a gaseous atmosphere – the Moon is not. As the young Earth began to cool, water vapour in its atmosphere condensed and rain fell, only to boil back. Eventually it began to rain in earnest – probably for several hundred years, with almost continuous lightning. The Earth's original atmosphere was very different from what it is today. It contained much water ($H_2O$), carbon dioxide ($CO_2$), methane ($CH_4$), nitrogen gas ($N_2$), ammonia ($NH_3$), some sulphur dioxide ($SO_2$) and traces of other gases, but no free oxygen.

To understand what happened next, we need to know that carbon atoms can form chains by sharing one or more electrons, like this:

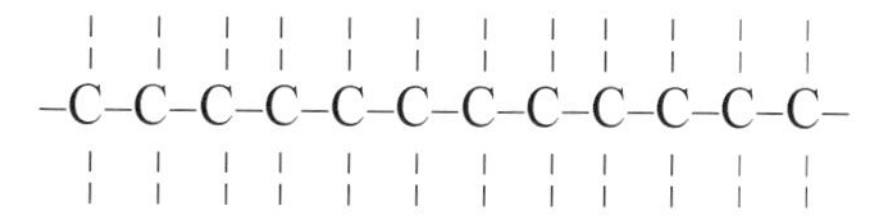

If hydrogen attaches to the spare hooks, it makes a *hydrocarbon*. Hydrocarbons are the basis of most fuels. For example, methane (natural gas) is $CH_4$, ethane $C_2H_6$, propane $C_3H_8$, and so on through to octane fuel (aviation spirit) $C_8H_{18}$ and nonane (petrol) $C_9H_{20}$, and paraffin and heavy oils to candle wax. They are called the *paraffin series*. The paraffins make good fuel because both carbon and hydrogen burn with oxygen to release energy; the heat given off by the chemical reaction causes the oxidised products, $CO_2$ and water vapour, to expand, and this expansion drives the pistons in a car engine or turbines in a jet. We will need to understand this elementary chemistry when we come to work out a way of measuring the energy-based wealth of nations in later chapters.

Because carbon atoms that are joined together in chains or rings are the chemical basis of all living things, carbon compounds are generally called *organic compounds*. Methane has a single carbon atom, not a ring or chain, so we call it an *inorganic compound*. In 1953, a Californian called Stanley Miller put what he believed to be the components of Earth's original atmosphere ($CH_4$, $NH_3$ and $H_2O$) into a flask and heated the mixture while passing electric sparks through it. The results were dramatic – the simple molecules reacted together to form a range of organic compounds, though none of them was a paraffin.[35] Repeated experiments yielded sugars, nucleobases and amino acids in a mixture of other compounds. We do not have to know the atomic structure of these classes of organic compounds, but merely think of sugars as rings of carbon atoms which include oxygen and hydrogen; some of them, such as sucrose (the white sugar on your table) are double rings. Amino acids are carbon chains with a *carboxyl group* (–COOH) at one end and an *amino group* ($–NH_2$) at the other. Nucleobases are rings of carbon and nitrogen.

The individual rings that are sugar molecules can join together to form chains of rings. For example, starch and cellulose are both chains of glucose (a sugar). Starch and cellulose are insoluble in water whereas sugars dissolve. Because the ratio of oxygen to hydrogen in these organic-based molecules is the same as in water, they are called *carbohydrates*.

Repeatedly heating and cooling, drying and wetting a soup of amino acids, as Miller did, causes some of them to join together in short chains called *polypeptides*. (Greek: *poly* = 'many' + *peptone* = 'a proteinaceous substance'). Though only about 20 amino acids are known to occur naturally, they can be assembled in any order to form chains of almost any length. Some amino acids have sulphur in them. Two sulphur atoms can link together by sharing an electron. If such a pair is on the same polypeptide chain, the molecule must have folded back onto itself, and the bond between the two sulphurs makes the folded polypeptide retain its shape. If a sulphur atom links up with another one in a different polypeptide, they will join the two polypeptide chains together. Cross-links of –S–S– are one way polypeptides can be folded or looped into an almost infinite variety of three-dimensional forms. When they do this, they are called *proteins*. There is no practical limit to the diversity of protein structure.

Proteins have two important properties. The first property is that they can form strong materials, such as hair, leather, silk, gristle and cartilage; such proteins are basically linear in shape, and they are vitally important in the structure of animal bodies. The second important property is that multiple folding of the amino acid chains allows proteins to form more compact, often globular, shapes that can hold other molecules in position while chemical changes occur in them. This vastly increases the rate at which reactions can take place. Any material that speeds up a chemical reaction is called a *catalyst*. Protein catalysts are called *enzymes* because they were first discovered in yeast (Greek: *en* = 'in' + *zume* ='leaven').

**********

How is this chemistry connected with the London questions? Because it is the basis of morality, and the first London question required a moral judgement. We can better understand the link between chemistry and morality with a thought-experiment. Consider this: it is uninteresting to know whether this particular hydrogen atom combined with that particular oxygen atom rather than another one, because all atoms of an element are identical in structure and behaviour. However, it *is* interesting to know that we can more easily digest glucose than fructose.

If we dissolve some natural glucose in water and shine a beam of polarised light through it, the plane of polarisation would be rotated to the right. An alternative name for glucose is dextrose. A solution of natural fructose would rotate polarised light to the left, so it is sometimes called laevulose. This difference in optical activity is caused by the two molecules, glucose (dextrose) and fructose (laevulose), being mirror images of each other, like our two hands, and they rotate the waves of light in opposite directions.[19]

Glucose, made artificially in the laboratory, is a mixture of half dextrose and half laevulose, and its solution does not rotate

[19] An engaging and clear explanation of optical isomerism can be found in Dorothy L Sayers's detective story *The Documents in the Case* (1930).

polarised light because the two effects cancel each other. If we add a dextrose-digesting enzyme (*dextrase*) to such a solution of artificially made glucose, molecules of dextrose and laevulose will collide with the dextrase molecules with equal frequency but only dextrose will combine with it and be digested. This is because dextrase requires any molecule, to which it can attach itself, to have a particular shape. Laevulose is the wrong shape: like a left hand in a right glove. Dextrose and laevulose clearly compete with each other over which shall combine with dextrase but only dextrose can fit the right-handed docking station on the very large globular protein that is the enzyme dextrase.

Having become attached to the dextrase enzyme, a dextrose molecule will begin to break down by drawing energy from its surroundings. Clearly this process would be greatly speeded up if an energy source were at hand. In fact, this is what happens. A special molecule, called *adenosine triphosphate* (ATP) provides this energy.[20] ATP has three phosphate ions attached to it in a line by energy-rich bonds (Figure 3.1).

An ATP molecule attaches to the dextrase enzyme when a dextrose molecule is also in position; the phosphate ion at the end of the row on the ATP breaks off and ATP becomes *adenosine diphosphate* (ADP). The energy in the last phosphate link is transferred to the dextrose molecule, which breaks up – it has begun to be digested.

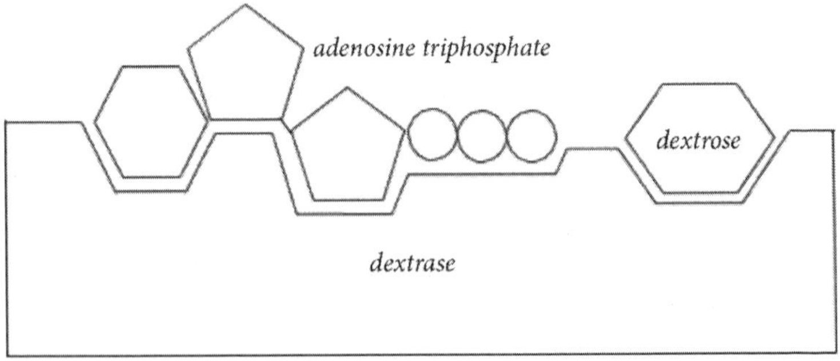

[20] Adenosine is a nucleotide that consists of a nucleobase (adenosine), a ribose sugar and three phosphate radicals.

**Figure 3.1** Stylised diagram showing how large organic molecules can fit together, and how this helps them to cooperate.

This is an example of how two different molecules, dextrose and laevulose, *compete* with each other, but the enzyme dextrase and the energy carrier ATP *cooperate* to achieve an end, which is the digestion of dextrose. Like the words 'compete' and 'cooperate', many people feel that the word 'achieve' is also loaded with human intention. No such special meaning is obligatory; indeed it hampers understanding in this case. In later chapters, we will look in more detail at how languages evolved and, being loaded with emotional innuendo, how unsuited they are to describing the physical world around us, especially chemistry.

**********

Proteins form the main structures of living tissues and, as enzymes, they control all vital processes. Proteins have one fundamental weakness: they cannot replicate themselves. To do that, they need to cooperate with a different class of organic compounds called nucleic acids. First, we need to understand that a phosphate is a group of oxygen and hydrogen atoms attached to a central phosphorus. For our purposes, the structure of phosphate groups is unimportant and we can represent them by circles, as shown in Figure 3.2. All we need to know is that they have two electron hooks. Ribose is a simple sugar consisting of a pentagon. Each pentagon can have several electron hooks. Ribose pentagons and phosphate groups can form chains in which they alternate:

**Figure 3.2** How phosphate groups link ribose pentagons together.

Nucleobases can join onto ribose rings so that they stick out from the ribose-phosphate chain. There are two kinds of nucleobase that interest us here: *pyrimidines* with one ring of carbon and nitrogen atoms, and *purines* with two. There are two kinds of pyrimidines, *cytosine* and *uracil*, and two purines, *adenine* and *guanine*. The shapes of these molecules are such that cytosine can link only with guanine, and uracil only with adenine; they do so, not with electron hooks, but with forces called *hydrogen bonds*.

We can represent cytosine and guanine, uracil and adenine with the letters C and G, and U and A, and Figure 3.3 shows how they attach to the chain of ribose and phosphate.

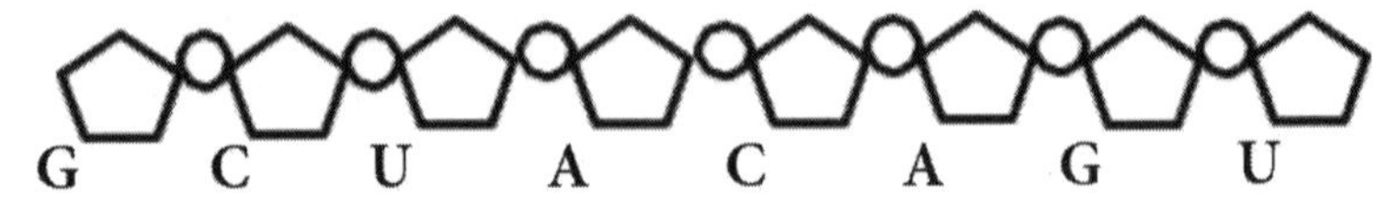


**Figure 3.3** Pyrimidines (C and U) and purines (A and G) attached to a ribose-phosphate chain.

The whole chain is called *ribose nucleic acid* (RNA), and it has three important properties: (1) it can act as an enzyme; (2) it can replicate itself; and (3) it can make proteins.

## 1. RNA can act as an enzyme

Just as a single polypeptide chain can double back on itself and hold the bends in place with sulphur-sulphur (–S-S–) bonds, so can a thread of RNA – if it is long enough. However, it holds its three-dimensional structure with hydrogen bonds between adenine and uracil or between cytosine and guanine, not with –S-S– bonds. Once an RNA thread has a more rigid three-dimensional structure, it can hold smaller molecules in place while changes occur on them, thus acting as an enzyme like the protein dextrase did.

## 2. RNA can replicate itself

To replicate the whole RNA molecule, each adenine in the chain links to a uracil molecule by means of three hydrogen bonds; each uracil joins to an adenine, each cytosine joins to a guanine, and each guanine joins to a cytosine by means of two hydrogen bonds. Now we have a chain with double nucleobases. Each of the exposed nucleobases now joins onto a ribose sugar and the ribose sugars link together with phosphates (Figure 3.4).

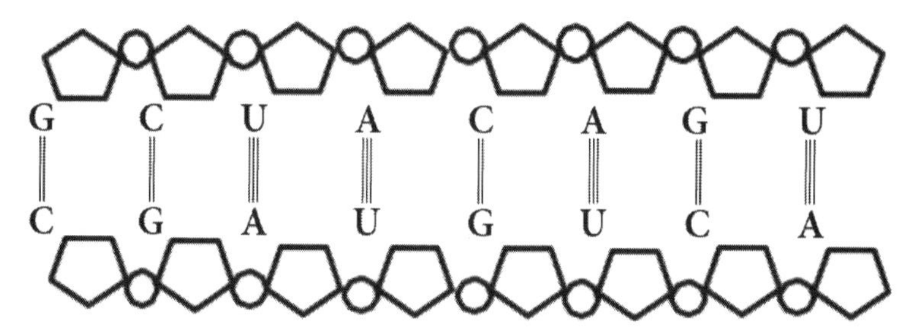


**Figure 3.4** How two ribose–phosphate chains can be linked together by hydrogen bonds between complementary pairs of purines and pyrimidines.

The nucleobases now separate and the two RNA chains pull apart; however, they are not identical. The newly formed chain is a mould of the original. This is because the nucleobases are not the originals, but partners of the originals. Now the mould-RNA molecule repeats the process and restores the original sequence of nucleotides – like making a cast from a mould. Replication is complete as soon as the cast separates from the mould.

## 3. RNA can make proteins

Shorter lengths of RNA fold back on themselves and twist into spirals. The shape is maintained by the coils being held together by hydrogen bonds between compatible nucleobases (A with U, C with G, G with C, or U with A). The lengths of twisted RNA are called *transfer RNA* or tRNA. They are represented by the ragged vertical oblongs in Figure 3.5. One end of tRNA is suitable for joining temporarily onto one of the 20 naturally occurring amino acids. At the other end of the spiral, which is the tip of the bend, there are three bare nucleobases, which can be any combination of A, C, G or U. Each combination is specific to the particular amino acid that links onto the other end of tRNA.

In Figure 3.5, tRNA picks up a particular amino acid, then waits for a space on a full length of genetic RNA (the row beginning GGA). Transfer RNA can link only with the correct sequence of bases that matches its own exposed triplet, and this is shown at (A and B).

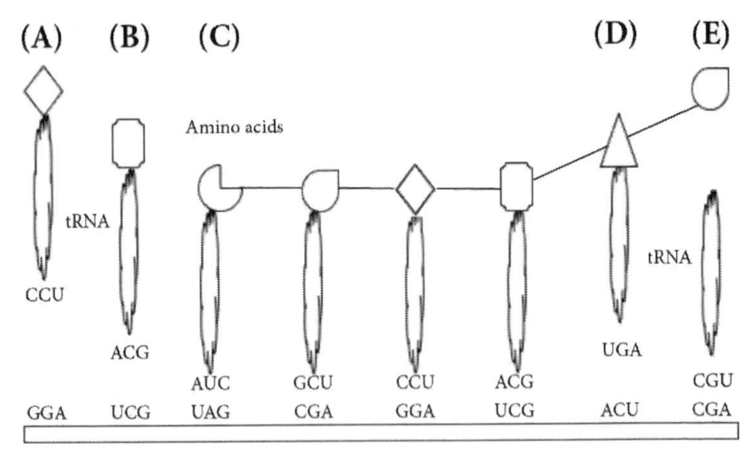


**Figure 3.5** Transfer RNA molecules carry amino acid molecules to a ribbon of genetic RNA so that they can be arranged in the correct order in a new polypeptide chain.

When tRNA finds its place on the genetic RNA, its amino acid joins onto the one immediately next to it at (C). This causes tRNA to release the amino acid it is carrying (D), allowing the amino acid to become part of a polypeptide chain. Transfer RNA now separates from genetic RNA and leaves the vicinity (E). It then joins up with another free amino acid of the right kind before returning to the genetic RNA and repeating the process.

Each of the 20 naturally occurring amino acids attaches to a specific tRNA that has a particular triplet of unpaired nucleic acids. For example the amino acid *phenylalanine* needs the triplet AAA, which has to find a place on the genetic RNA chain where UUU occurs. The amino acid *alanine* has a triplet consisting of GCA, so it needs to find a site where CGU occurs, and so on.

**********

I have described this process in some detail because it is the mechanism behind probably the most astonishing fact in all of biology, if not in all of science, which is that every single living thing ever discovered, from viruses, through to bacteria, cabbages, birds of paradise, whales, trees and us uses exactly the same three-letter code for the 20 amino acids. This is compelling evidence that life evolved on Earth only once.

Clearly, the order of amino acids on the newly formed polypeptide chain is controlled by the sequence of nucleobases in the genetic RNA. The eventual shape of the newly formed protein is determined by the distribution of sulphur atoms on the polypeptide chain. As already described, this fixes the three-dimensional shape of the protein.

The effectiveness of the process is achieved by a balance between competition and cooperation: nucleobases compete with each other for places on the RNA chain; amino acids compete with each other for places on the polypeptide chain; genetic RNA molecules cooperate with moulds and casts in their replication and with tRNA to synthesise proteins. More than that, the process is also an intimate cooperation between two kinds of chemical *property* (structural and replicative), as well as between two kinds of chemical *class*, (RNA and protein), and this cooperation forms the basis of life.

The processes of both replication and protein synthesis allow room for error – there can be mistakes in how amino acids are assembled to make proteins, and there can also be mistakes in how RNA replicates. The possibilities of such errors add another essential ingredient of life – variation. So long as there is variability, there is room for improvement: without it there can be none. This fundamental property of mere chemicals lays the pattern for all of evolution; and evolution gave us our social systems. This is why variety and error are so important, and why those who would impose on others a fixed pattern of thought are so dangerous, and in the long term futilely ineffective. Nowhere is this more evident than in the present crisis of Islam.[36]

**********

A soup of structural proteins, enzymes and RNA is not a living organism. To be that, an organised structure is essential.

Closely related to the paraffins are a group of organic compounds called *lipids*. Lipids differ from paraffins in that, instead of having a hydrogen atom at each end of the carbon chain, one of them is replaced with a carboxyl group –COOH, which we have met before in amino acids. Figure 3.6 shows its structure in more detail.

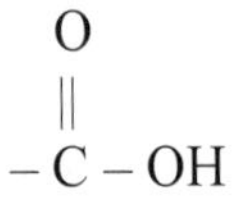

O<br>
||<br>
– C – OH

**Figure 3.6** The structure of the carboxyl group.

Like when it is exposed at the end of a polypeptide, the carboxyl group can ionise when in contact with water by losing a hydrogen ion (Figure 3.7).

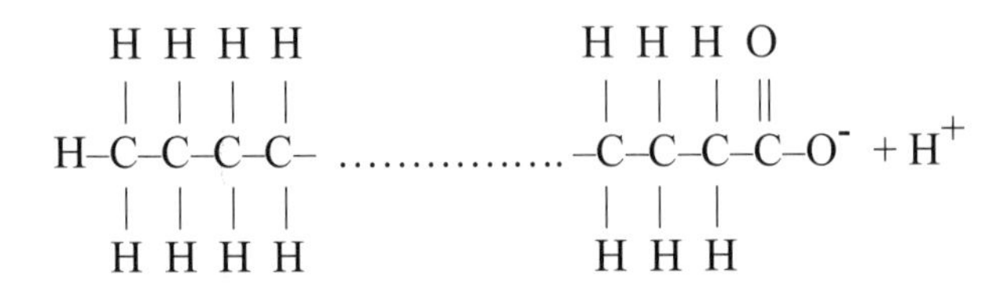


**Figure 3.7** Ionisation of the carboxyl group.

We can write the whole carboxyl molecule as the symbol $–Q^-$.

If you put a drop of liquid lipid onto the surface of still water, it will spread out to form a circular patch because all the molecules will stand on their free $–Q^-$ heads with their oily tails in the air. We can think of the heads as being water-loving and the oily tails as being water-repelling. If you shake the lipid and water together, droplets form because the lipid molecules surround drops of water in a double layer with their oily tails end-to-end, and the water-loving $–Q^-$ heads of the inner layer facing inwards towards the drop of water they surround, and the outer $–Q^-$ heads outwards towards the water that surrounds them, as shown in Figure 3.8.

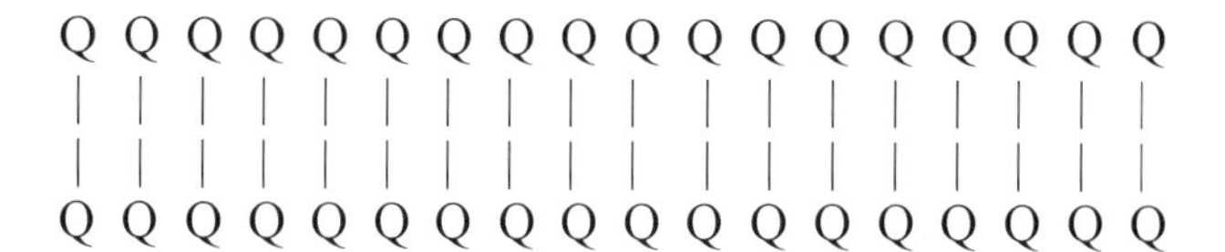

**Figure 3.8** Lipid molecules arrange themselves in a double layer with their water-loving ends outwards and their water-repelling tails inwards.

As they form droplets, the double layer of lipids bends into a sphere. In effect, lipids readily form two-layered vesicles around a drop of water. Almost any small molecule can pass through a double lipid membrane because it is permeable and also very fragile; notice the droplets in shaken oil and water coalescing as they rise to the surface.

The coverings of proteins can link to the water-loving ends of the lipid molecules, inside and outside the double lipid layer, and so make it physically stronger. Pores through the four-layered (protein-lipid-lipid-protein) membrane can control the size of molecule that can pass to and fro. This makes the membrane selectively permeable; that is to say, the pores will allow certain molecules, but not others, to pass through them (see Figure 3.9).

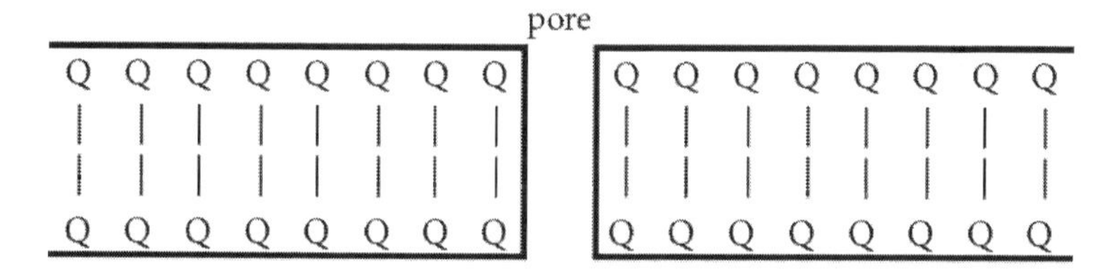


**Figure 3.9** A selectively permeable membrane.

These true membranes surround RNA in a watery solution (called the *cytoplasm*) and retain materials needed for RNA to make proteins and to replicate. The whole structure is called a *cell.*

Enzymes can become attached to the membrane and, if they do so in the sequence of procedures they catalyse, molecules they are handling can move from one enzyme to another with enhanced efficiency. In addition, the chemical reactions of RNA become even more efficient because the raw materials required by their processes can be selectively absorbed through the cell membrane and waste products expelled.

**********

We call a length of RNA that codes for a single protein, a *gene*. Genes will inevitably compete with one another, not only for amino acids as they manufacture the proteins they code for, but also for nucleotides when they come to replicate themselves. However, even though each gene codes for a different protein, the two proteins may combine to form a functional structure in the living cell. In this case, there is no point in the two proteins being produced in anything but equal quantities, nor is there any advantage to the cell if one of them becomes more abundant than the other. Indeed, there is a disadvantage if this happens because scarce and valuable materials and energy would be wasted. Therefore, it would be an advantage to the economy of the cell if a mechanism evolved that prevented either of these two genes from becoming more abundant than the other. The solution to this problem is obvious: simply joining the two genes together, end-to-end, ensures that they replicate at the same time.

Up to a point, it is more efficient for the synthesis of proteins if the replication of all those genes whose functions are coordinated are linked in this way. Such strings of genes were identified early in the study of cells by means of dyes that coloured them; accordingly, the strings were called *chromosomes* (Greek: *khroma* = 'coloured' + soma = 'body').

The interactions between RNA and proteins form series of processes that increase the longer-term efficiency of RNA replication. Both these classes of chemical compound can act as catalysts. When they do so to promote the efficiency of their own reactions, they are called *autocatalysts*. It is the escalation of autocatalysis into multidimensional webs that we call life.[37]

**********

Chemical reactions occurring near the chromosome are likely to be concerned with replicating the chromosome or conveying information on transfer RNA to the rest of the cytoplasm. Reactions further from the chromosome would be more likely to be involved in manufacturing proteins and releasing energy. Therefore, there would be an advantage to the cell as a whole, if these two groups of processes were separated by another selectively permeable membrane. So a new membrane evolved around the RNA – the *nuclear membrane*; the special cytoplasm contained within it is called the *nucleoplasm*.

Eventually, a plethora of membranous organelles evolved in the cytoplasm, that is between the nuclear and cell membranes, each organelle performing a specific suite of functions. Some of these organelles contain RNA while others do not. Any RNA in these organelles is distinct from that of the nucleus. This suggests that those organelles which contain RNA were once independent living things that have become incorporated into the evolving cell. One of these types of RNA-containing organelles specialises in releasing energy, and they are called *mitochondria* (singular *mitochondrion*). Another type contains plates of a green pigment called *chlorophyll*, and they are called *chloroplasts*. Chloroplasts are found in green plants and trap light energy. This combining of two or three distinct species of living organism into one organism is cooperation at a much higher level.

As we have seen, an RNA thread can replicate itself by assembling molecules of nucleotides, ribose sugar and phosphate and arranging them in order. During the process of replication, the template briefly created a double thread of RNA that then separated and repeated the process to make a copy of the original. Permanently doubling the RNA thread gives a simple and effective way of ensuring that mistakes do not occur too often when the chain replicates itself. This is achieved by each of the nucleotide pair insisting on the right partner.

When the RNA thread that is a chromosome doubles permanently, ribose sugar is not quite the right molecule for maximum efficiency. Double chains work better if ribose molecules lose an atom of oxygen, and this makes another kind of sugar called *deoxyribose.* Plant and animal cells use this deoxyribose sugar in their genes. So their chromosomes are made of deoxyribose nucleic acid or DNA.

One other small complication is that DNA chains do not use uracil, but a similar pyrimidine called *thiamine*. Thiamine pairs with adenine, just as uracil did. However, plant and animal cells still use RNA for picking up amino acids, and also for reading DNA. The doubled DNA chain looks rather like a stepladder with the deoxyribose sugars and phosphates forming the rails and the nucleic acid pairs forming the rungs (Figure 3.10).

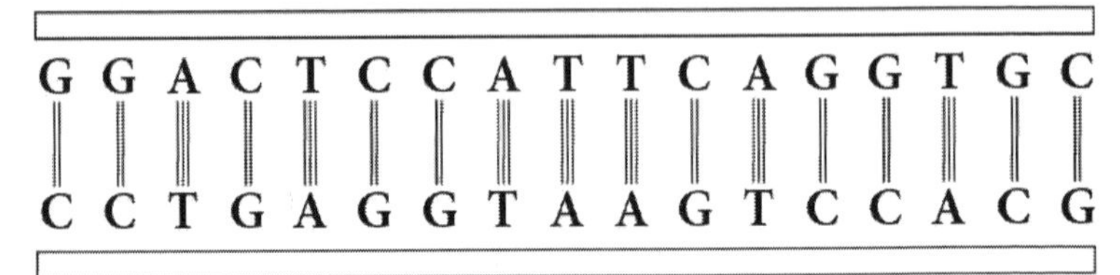


**Figure 3.10** The doubled DNA chain

The angles between the deoxyribose sugars and the phosphate molecules that link them twist the ladder into a spiral or helix.

Before the sequence of nucleic acids that form the rungs of the twisted ladder can be read, the double DNA molecule must be untwisted and the rungs opened in the middle to expose the nucleobases. A very large mass of RNA, which is convoluted into a solid mass to act like an enzyme, latches onto the DNA chromosome to straighten and open it. When this happens, fresh nucleobases join onto their exposed partners: guanine with cytosine and *vice versa*, thiamine with adenine, and adenine with uracil. Yes, uracil not thiamine because the sugar that forms the thread with phosphate is ribose not deoxyribose, and so the new thread is RNA. This new thread of RNA carries the information of the nucleic acid sequence on the genes, so it is called *messenger RNA* (mRNA). (I called it 'genetic RNA' in Figure 3.5.) Messenger RNA now passes through the nuclear membrane and assembles polypeptides using tRNA as described previously.

If any entity can behave in one of two ways, it is effectively making a choice; in other words, it makes a decision. The decision need not necessarily be a conscious action, nor need it be a cascade of neural events (which are physico-chemical anyway) in a human or animal brain. In this sense, a single molecule can decide whether to accept or reject the shape of another molecule. This points strongly towards competition and cooperation being properties of chemical reactions between molecules. Such choices become observable only when molecules become recognisably different. Once we recognise this, the rest of the story is straightforward. I tell it to show that there is no known break in the evolution of complexity in moral choices between that shown at the first appearance of molecules and, say, the agonising of a judge. The mechanistic bases of morality are prosaic and understandable – there is no need to invoke divine or other supernatural intervention.

Now we need to look at the actual mechanism of evolution; only then will its story make sense.

# Chapter 4

## EVOLUTION: THE MECHANISM & THE STORY

The definition of evolution as 'the survival of the fittest' is wrong. It was coined by Herbert Spencer in his *Principles of Biology* (1864) and once shed a little watery light on a new subject. Yet, it is still trotted out as if it were a newly discovered truth.

First, 'fittest' means 'those most fitted', and has little to do with athletic fitness. Some individual animals of spectacular size and vigour are clearly good at surviving because they live to old age – they are 'fit'. However, some of these spectacular individuals owe their magnificence to the fact that they are sterile. This means that they were untroubled by sexual drives, and so concentrated their energies on feeding, growing and escaping predators or rough weather. They are obviously supremely well fitted to the environments in which they live but, though they survive brilliantly, they are totally 'unfit' evolutionarily speaking, because they leave no offspring.

Success in evolution is measured by reproductive advantage, nothing more. The individual that leaves more offspring than another is fitter in the evolutionary sense. But this is a tautology:[38] if it is fitter, it leaves more offspring – same thing and it does not get us anywhere. I shall simply drop the word 'fit' from our thinking about evolution. Also, I shall use animals as examples, though the principles I discuss are equally true of all living things.

**********

Now to the mechanism of evolution: we notice that eggs and young animals are more numerous than adults, if only at certain times of the year. In other words, living things multiply. Why, then, do populations of species not go on increasing indefinitely? In fact, though they may increase for a period, populations eventually remain stable or decline – which is to say that some individuals die before they can breed.

We also notice that no two individuals are exactly alike; even identical twins differ because each of the pair is exposed to different environmental influences, even in the womb. This variation between individuals within a species is not just in outward appearance but also in rate of growth, longevity, vigour and fecundity, which is that some individuals will leave more offspring than others. Since there is a genetic link between generations, successful parents will tend to pass on to their offspring the adaptations that led to their success.

*That* is the theory of evolution, nothing more. We can summarise it in six words: *multiplication with inheritance, variation and selection*. Its beauty is its simplicity, its elegance and its robustness. It has operated efficiently throughout the long history of the Earth, even among pre-life chemical compounds. No theory has been so rigorously investigated or hostilely attacked, yet its core remains as reasonable as it was when Charles Darwin and Alfred Wallace[21] first formulated it.

We also use the simple process of evolution to create new breeds of domestic animals. We do this by simply choosing which animals in our flock, herd, kennel or loft we shall allow to breed and which ones we shall not. We base our choice on the physical, physiological or behavioural characters we desire or reject.

Darwin devoted the first chapter of *On the Origin of Species* (1859) to his study of how varieties of domestic animals were improved, and he corresponded widely with breeders. Being aware of public sensibilities, the breeders did not

[21] Wallace is nearly always referred to as Alfred Russel Wallace, though I can find no evidence that he used his middle name regularly. There is a theory that the single 'l' was a misspelling by the Welsh registrar of births. I suspect that if Darwin's middle name had been registered as 'Robertt', it would be as frequently included to display the learning of those who used it. As I shall argue, display is one of the most powerful and least talked about drives in human behaviour, and has a profound influence on the present population crisis.

emphasise the fact that they killed most of the animals they did not intend to breed from: they *selected* the stock from which they bred subsequent generations. Darwin was disgusted by cruelty to animals,[39] so he too concentrated on selecting desirable stock rather than eliminating the undesirable, even though those undesirable individuals are also selected but in a negative sense. Living things in their natural environment seldom have such consideration or the mercy of a quick death. And this highlights an aspect that we shun in our sanitised and essentially urban world: evolution is not influenced by birth rate; it works on pre-reproductive death rate.[22]

Evolution operates in every system in which there is multiplication with inheritance, variation and selection. For example, to improve computer programs, we can instruct one to carry out the following process: make 1000 copies of itself – all slightly and randomly different – test each new program for its efficiency in performing a defined task, choose the most efficient one, delete the other 999, repeat these steps thousands of times starting with the most efficient program each time.

Because the variations are random, the vast majority of the programs are not as efficient as the original. But it may happen that, without any forethought, chance produces one tiny improvement. Testing identifies that improvement and selection preserves it. Humanely deleting the other 999 programs saves space on our hard drive so allows room for the more successful versions. In this way, we slowly accumulate improvements in efficiency at performing a particular task. The mechanism of evolution is no more complicated than that.

**********

[22] This is not strictly true, though it is in terms of sheer numbers. Birth rate certainly affects variability.

## A historical digression[23]

Until the early 17th Century, most intellectual energy in Europe was directed towards studying Christian or classical literature. Interest in the natural world really began with the Enlightenment, a period when bright minds laid the foundations of modern understanding. During the 18th Century, travellers collected volumes of factual information about the world, including many specimens of plants and animals. Scientists, or 'natural philosophers' as they then called themselves, studied and classified these specimens and arranged them in museums. Most museum curators put rocks and minerals (for example, apatite, haematite, quartz) in different cabinets from the remains of living things, and kept plants separately from animals. Then they put hairy animals together and kept them apart from those with feathers. In doing this, they noticed that there were clearly groups of living things that, though they varied slightly among themselves, *looked* distinctly different from other such groups, particularly when they all came from one region. The curators called these groups *species* (Latin = 'appearance', from *specere* = 'to look').

Curators described each species and laboriously referred to them by these descriptions. They were striving to understand the mind of God by discovering the basic units of creation – the 'type' of each species. Eventually they condensed the species descriptions into a short paragraph. The next advance was when the Swedish physician and naturalist Carl Linnaeus (1707–1778) reduced the descriptive paragraphs to two words – the two-name system that is now used internationally to name species. Linnaeus was the son of a priest and had no reason to challenge the Biblical account of the Creation as an adequate explanation for the diversity of living things on Earth.[40] However, during the 19th Century, scepticism about the role of God as creator grew among intellectuals.

[23] There is a plethora of biographies on Charles Darwin. I think the best is that written by Browne (1995, 2002). More than 7,500 of Darwin's extant 14,000 letters are available online from the Darwin Correspondence Project website (http://www.darwinproject.ac.uk (October 2015).

Charles Darwin was the son of a wealthy English doctor. He was born in 1809 and grew into a dull boy who was stirred into activity only by frivolity and shooting, so his father described him.[24] He went to the University of Edinburgh to read medicine so that he could follow in his father's practice but, disgusted by the dissecting room, he failed to complete the first year. Almost in desperation, his father sent him to Cambridge to read theology. But young Darwin fell under the spell of botany and geology, and became a passionate naturalist with an inordinate fondness for beetles.[25] By luck and a good deal of family influence, Darwin got himself the position of captain's companion on HMS *Beagle*, a naval survey ship that was to journey round the world.

In the English navy of 1831 – the year *Beagle* sailed – ships' captains lived aloof from their officers, dining apart and speaking to them only to give orders. This enforced isolation occasionally affected their sanity; indeed, having struggled with depression for many years, *Beagle's* captain, Robert FitzRoy, eventually took his own life. To reduce this risk, captains sometimes travelled with men of equal social status as companions. The official naturalist on *Beagle's* voyage was Robert McCormick, who combined these duties with those of ship's surgeon. Having been disappointed in his expectations of carrying out his natural history pursuits, McCormick left the ship four months after sailing. He had some reason to because FitzRoy favoured Darwin's natural history ambitions over those of the official naturalist; there was also a clash of personalities.[41]

Darwin's passion for natural history became an obsession; he developed enormous energy, kept copious notes on everything he saw – particularly the geology – discovered new species,

---

[24] Bowlby (1990: p. 70) *Charles Darwin: a New Life*. His father's words were: 'You care for nothing but shooting, dogs and rat-catching, and you will be a disgrace to yourself and all your family.'

[25] Desmond & Moore (1992: p. 57); for an interesting correspondence on the phrase that was made famous by J. B. S. Haldane in response to a question about the interests of God, see Williamson (1992: p. 14) and Cain (1993a: p. 13).

wrote hundreds of letters to learned and influential people, collected and sent home specimens by the trunk load, and was already famous in London before he reached the Galápagos Islands, aged 26. *Beagle's* voyage took five years. About a year after he returned,[42] Darwin began writing a secret notebook in which he jotted his thoughts – mostly about the mutability of species.

For three years after his return, Darwin worked himself to exhaustion and illness, compiling a report of his journey and having his specimens identified or named by specialists. His wealthy father gave him an allowance of £400 a year, a large sum in those days. At 31 he married his first cousin Emma Wedgwood. The couple lived in London where Darwin could attend meetings and consult people. But they loathed city life and moved to Kent in 1842 where they eventually had 10 children.

Today, it is difficult to imagine the certainty with which most people believed the Biblical account of the creation of the world, life and Man being created in the image of God. Refuting the existence of God, in those days, was similar in offensiveness as saying today that human life is the cheapest commodity on Earth. Central to the belief in divine creation was the conviction that species could not change one into another. Had such mutability been accepted, the whole concept would have become open to criticism. Nevertheless, some intellectuals suspected that species did change. They even went so far as to discuss this heresy, but they had no mechanism to explain how it worked. It is not clear whether Darwin's understanding occurred in a moment of inspiration, or emerged slowly from his almost continuous thinking about the matter. However it happened, his was one of the greatest insights in all of science – he had concluded that the mechanism of evolution was *multiplication with inheritance*, *variation and selection*.

Emma was deeply religious, and Darwin was appalled at the thought of what publishing his ideas would do to her faith. She was equally horrified at what these ideas would do to his immortal soul. She simply could not bear the thought of him writhing in Hell for all eternity, and that they would not meet again in Heaven. In addition, many of the men Darwin consulted and respected were fiercely opposed to any argument against creationism. So, for the next 15 years Darwin gathered evidence for *and against* his theory. Upon analysis, he found that the evidence against it did not stand up.

In 1858, Darwin received a letter from Alfred Wallace, a naturalist working in what is now the Indonesian Archipelago. In the letter, Wallace outlined a possible mechanism for evolution – multiplication with inheritance, variation and selection – which had occurred to him in a flash of fevered inspiration. Darwin was in despair, all his originality and decades of hard work had gone because priority in publishing was then, and still is, paramount. He consulted with friends who supported his theory and they reminded him that he had 'published' his theory by telling them of it. As a solution to the quandary, they recommended that he and Wallace publish a joint paper at a meeting of the Linnean Society of London. Neither man attended the meeting. Wallace was in the Far East and did not even know about it, and Darwin was burying his baby son. The paper excited some interest among the Fellows of the Society who heard it, but the President prevented discussion to keep the meeting short. The next year Darwin published his book, *On the Origin of Species by Means of Natural Selection*. It was an instant bestseller. It has never been out of print, and changed forever the way we see the world, including ourselves.

Though Darwin's theory of the mechanism of evolution has received 150 years of sustained investigation by hostile researchers, it remains the best explanation we have for the facts we observe. No theory can be *proved*; it can only be *disproved* by new evidence and accurate thought.[26] Darwin would have been the first to welcome another theory to replace his, as would any modern scientist of intellectual integrity, so long as it met the rigorous conditions of peer review.

[26] An important contribution to thought by Karl Popper. See Popper (1959) and Magee (1985: p. 22).

Although the search for an alternative theory has so far been unsuccessful, it goes on.

Darwin's mechanism of evolution does not depend on any particular way of reproduction, nor on any method of variation. Whether species increase in numbers by spores, binary fission, elaborate sexual ritual or clicking a mouse button is unimportant. Nor does it matter whether variation is caused by genetic mutation, recombination of inheritable particles, blending or characters acquired during an individual's lifetime. However, the theory does insist that there is a system by which information is passed from generation to generation. The fact that Darwin was completely wrong in his understanding of inheritance, in no way undermines the theory.

Like other biologists of his time, Darwin thought that each part of the body produced 'gemmules' that migrated to the reproductive organs and became incorporated into the sex cells: pollen or ovules in plants, sperm or eggs in animals. These gemmules carried with them the characters of the organ that produced them as it was when they left it. So, if an organ changed during the lifetime of the individual, these features would also be transmitted. Clearly, if an organ had not developed or had been lost, it could not produce gemmules, so that feature would be missing in the offspring. For example, cutting off the tails of mice should result in tailless baby mice. It is strange that this easy experiment was never carried out until much later.

Jean-Baptiste Lamarck[27] (1744–1829) was the youngest of 11 children of an impoverished French aristocratic family. He was always interested in natural history and collected shells avidly. After most of his career studying plants, he became professor of insects and worms at the National Museum of Natural History in Paris. Insects and worms? Aristotle had classified animals into: mammals, birds, reptiles, fish, insects and worms, and this system stood unchanged for more than

---

[27] Much of what follows is from Gould (1980: ch. 7) and Gould (2000: p. 119).

2000 years. Even the great revisions of the 18th Century persisted with 'insects and worms' as a ragbag for complex animals without backbones. We owe Lamarck for the word *invertebrate* for this heterogeneous assemblage, and for the names of the main invertebrate categories that we use today. He also gave the term *biology* to that great discipline.

Looking back on the range of complexity among living things, Lamarck thought that there was a *deliberate* progression from simpler to higher forms. He envisaged a continuous force that drove evolution to more and more complex organisation, though he did not even speculate what the nature of that force might be. This accounted, he thought, for the apparent ladder of evolution from single-celled organisms to mankind. He called this force *the force of increasing complexity*.

Lamarck also suggested that the *effort of striving* to achieve a function modified the gemmules to that end. His often-quoted example was the long neck of the giraffe which evolved because giraffes continually made the effort of stretching upwards to reach the higher leaves of trees on which they fed; shorter-necked animals having already eaten the lower ones.

Lamarck was very much aware of the huge diversity of nature and particularly of the vast array of species that are not near the ancestral line that led to us. To account for these branches, he suggested *a force of circumstantial influences*. Thus, acacia leaves being high in trees influenced giraffes to make the effort of stretching their necks. Forces of circumstantial influence operated alongside the central force of increasing complexity. We now know that no such forces exist in gene-based evolution.

In contrast to Lamarck's theory, Darwin argued that more young giraffes were born than survived. By random chance some of them grew longer necks than others and this gave them an advantage when it came to feeding or seeing distant predators. Therefore, they were more likely to survive and breed than those with shorter necks. In other words, they had been positively selected. Having been selected, they then passed on the tendency to grow longer necks to their offspring. We now know that striving does not alter heritable particles, so Lamarck's theory is exploded. Darwinian evolution has nothing to say about how inheritance works, so remains centre stage. Nor has it anything to say about what sort of entity is evolving: *multiplication with inheritance*, *variation* and *selection* applies perfectly well to genes, sex cells, individuals, families, populations, species, ideas, businesses, motor cars or computer programs. But Lamarckian evolution would only work for whole living individual organisms.

It is a sad reflection on much school teaching that Lamarck, a truly great and innovative man, should be remembered only for one subsidiary idea he got wrong, and not for the brilliant light he shed on the central discipline of biology. Lamarck survived the horrors of the French revolution, changing his title to 'citizen', and died blind and in poverty.

**********

So much for the *mechanism* of evolution. Different from it is the *story of the changes* that living things experienced as one species evolved into another. The earlier changes are more difficult for us to relate to than more recent ones, even though they laid down the fundamental body designs that characterise the main types of plants and animals.

From time to time, the communications media announce that scientists have created life. The nearest to such an achievement so far was in 2010 when the American biochemist, geneticist and entrepreneur Craig Venter claimed to have created artificial life by inserting a chromosome he had assembled into an existing cell from which the original chromosome had been removed. That cell was 'living' in that it showed all the characteristics of life except being able to reproduce. That is to say, it could absorb nutrients, release energy, excrete waste and respond to changes in its surroundings. Venter's experiment restored its ability to reproduce.

His achievement was like designing a nuclear submarine from scraps of pre-existing plans, and taking the design to an up and running engineering industry to build it. Had Venter also constructed a cell from inorganic materials – like creating mining, refining and engineering industries from raw materials – his achievement would have laid better claim to having created artificial life. At the molecular level, the living cell is, in fact, far more complicated than a mere human artefact, such as an industry capable of building a nuclear submarine.

No one has been able to create life. A few decades of sporadic research does not give the same probability of success as millions of years of continuous experiment. It is more parsimonious to think that life began by accidental collisions between molecules over vast periods of time than to invoke the intervention of a god. It is just possible that life on Earth was seeded by a meteorite but to assert that merely transfers the question elsewhere.

There was no moment at which chemistry became biology, no sudden spark after which we could say, 'Now there is life'. Much more likely is that there was a long sequence of tiny alterations that involved replicative systems of chemicals which could catalyse the reactions they took part in – they were autocatalytic. Eventually, the sequence of alterations culminated in the behaviour of competing and cooperating molecules that fulfil our modern definition of life.[28]

From that wholly artificial point, the story unrolls in a pageant of grandeur that must be rare, if not unique, in our universe. Obviously, there is no space in a book like this to do more than list, rather than explain, the most likely steps we have taken in our journey to where we are: masters of our planet and our fate. You can find descriptions and illustrations of the animals I mention on the World Wide Web or in books. Perhaps the best account of animal evolution is given by Richard Dawkins in his *The Ancestor's Tale: a Pilgrimage to*

---

[28] Pross (2012) gives an up to date account.

*the Dawn of Life* (2004), and I commend it.[29] In short, the evidence for organic evolution flaring across the Earth to create the magnificence we can still see in a few, fast-disappearing places, is soundly based on testable observations and logical argument, rather than on what people and ancient texts claim.

**********

As mentioned in Chapter 3, mitochondria, the energy-releasing organelles that float freely in the cytoplasm, contain their own packets of DNA and RNA and reproduce by dividing. This indicates that an ancestral cell ingested a living bacterium but did not digest it, and used the bacterium's special chemical talent for releasing energy in exchange for nutrients and shelter; this is a relationship called *symbiosis*. Because the egg is larger than the sperm, all the mitochondria in our bodies come from our mothers, and none are derived from the sperm that fertilised the eggs that became us. That is why it is possible to trace the evolution of the human female line back to one hypothetical woman popularly called 'Mitochondrial Eve'.

Another type of organelle within a cell is the chloroplast, which is found in plant cells. Because, like the mitochondrion, the chloroplast also contains its own DNA, it too was once a free-living organism, in this case a photosynthetic bacterium that was ingested and retained alive to continue its function of trapping light energy and converting it into a form that could be used in physiological processes.

Here are two more examples of cooperation at an early stage in our evolution: both host and lodger obtaining some advantage from the relationship. There are many known instances in which animals have symbiotic algae living inside their cells. These algae need light; if light fails and they no longer supply their animal host with sugar, the animal simply digests the lodger. I see no moral difference between this and a bank foreclosing on a business that fails to pay its interest.

---

[29] A striking pictorial version of the evolutionary story can be found in Palmer (2009).

The Earth formed about 4.5 billion years ago. As I explained in Chapter 3, the appearance of life took a long time. All we can say is that the first simple cells had evolved by about 3.5 billion years ago, and that the first higher cells with mitochondria were present 1 billion years ago. Since these numbers of years are almost unimaginable, it is clearer to scale Earth's history into 1 year, so that 1 day represents 12.3 million years. On this scale, simple cells appear on 2 May, and higher cells on 11 October. The more conspicuous forms of animals as we know them evolved suddenly on 20 November. Dinosaurs ruled the Earth from 15 December to Christmas Day when mammals and birds replaced them. We human beings walked onto the stage at 9.30 pm of 31 December. Our recorded history stretches back to 5 minutes before midnight on that last day.[43]

To start with, the first single-celled organisms simply absorbed energy-rich molecules from the sea in which they lived. This source of food soon began to run out, so some cells obtained a competitive advantage by excreting enzymes onto insoluble organic compounds they encountered, and breaking them down into molecules that were small enough to absorb. It did not matter to the feeding cell whether its food was dead or alive. The distinction between scavenger and predator was strictly practical in that dead things were less resentful than living ones. Predators attacked their prey either by excreting enzymes onto it, or by engulfing it. In the first case, such predatory organisms remained small and evolved into parasites that cause diseases. To engulf whole cells, there is an advantage in being large. Large size not only enables a predator to attack more prey, but is also a defence against such attacks. So began the great arms race between predator and prey that produced large mammals and dinosaurs. The human body is fourth largest, after the three gorilla subspecies, of the 479 surviving primate species.[44]

The structure of the living cell is not suited to increasing size. To become larger, an organism has to become many-celled. This happened in two ways: individual cells cooperatively accreted together, as they do in sponges and slime moulds, or a single cell divided and the daughter cells remained together as they grew. Higher animals opted for the second method, largely because it is easier to ensure that the genetic instructions in all cooperating cells are identical.[30] A mutation in a body cell can lead to cooperation between it and other cells breaking down because it is no longer genetically identical to them. The result is often disorganised growth of the mutant cell, and is called cancer.

**********

Comparing two living species of animal shows a likely route to organisms becoming many-celled. One, *Paramecium*, is a single-celled animal though it has two nuclei, and is up to a third of a millimetre long. The other, *Symsagittifera*, is about 15 mm long and many-celled, though some of the cells that line its gut are incompletely separated. Both animals are aquatic and move by waving fine hairs that cover their bodies. Both feed by taking food into the body: *Paramecium* directly into the cell and *Symsagittifera* into an opening in the body; then, the cells lining the gut pinch off bits of food and engulf them. Though there are other similarities, they do not suggest that *Symsagittifera* evolved from *Paramecium*; but it does allow for a common ancestor that was probably more like *Paramecium*.

A *Symsagittifera*-like creature makes a good hypothetical ancestor for most many-celled animals. That ancestor could have adopted a sedentary way of life and evolved into sea anemones, jellyfish and corals. Jellyfish, which are like upside-down, free-swimming sea anemones, may have been the first free-swimming, many-celled predators, sweeping the sea floor with stinging tentacles to catch animals living there.

If that is a realistic scenario, other stocks of the *Symsagittifera*-like ancestral animal had various ways of escaping such predators. One form could have burrowed into

---

[30] An interesting corollary is that slime moulds gather together to reproduce, and they collectively decide which cells shall pass on the genes they carry. Democracy?

the silt of the sea floor like modern lugworms and other segmented worms. Other stocks could have secreted shells like modern molluscs or sea urchins. Another could have burrowed into the stems of seaweeds like modern roundworms. And yet others could have remained flat and hidden under stones like modern *Symsagittifera* and other flatworms. In fact, each of these groups of animals has a fundamentally different body design. Which category of animals mentioned would you, perhaps as a non-zoologist, think was the most likely candidate for our ancestor?

Sea urchins and their relatives.

Sea urchins, starfish and their relatives are collectively called *echinoderms* (Greek: *echinos* = ' hedgehog' + *derma* = ' skin', because that is the appearance they generally have). Echinoderms are radially symmetrical like a jam jar or star, and their symmetry is pentagonal. This is because a pentagon is a compromise between conflicting evolutionary requirements. A triangle is the figure with the fewest straight sides to enclose space but it wastes a lot of it in awkwardly sharp corners. A square and a hexagon both have angles opposite each other and this presents planes of structural weakness; but a pentagon always has a full plate opposite each angle.[45] Maybe this feature played a part in saving most of the structure of the US military headquarters at the Pentagon in the outrages of 11 September 2001.

Echinoderms and all animals with backbones (including us) have one kind of embryonic development, and all other animals have different kinds. This is strong evidence that we are more closely related to echinoderms than to any other invertebrate. There is another group of animals, which appears intermediate between echinoderms and vertebrates, and they are called *protochordates*. Protochordates include acorn worms, lancets and sea squirts, and the early stages of their planktonic larvae are remarkably similar to those of echinoderms. Adult sea squirts are marine filter-feeders that stick to rocks in dark crevices, and their later larval stages are like tadpoles. This larval form evolved because it has to swim accurately to find a place where it could successfully develop into an adult. From several possible ways of achieving precise locomotion, evolution selected a tail that waggled from side to side.

Adult sea squirts have gills and reproductive organs, while their larvae have segmented muscles pulling against a flexible rod that eventually evolved into the precursor of the vertebrate spine. Only tiny genetic adjustments are needed to change the settings of genes that initiate particular developmental stages; so it is evolutionarily not difficult for gills and reproductive organs to appear precociously in the larva. The axolotl is a large Mexican salamander, and it retains larval gills as an adult. When this happens in an individual – as in the axolotl – it is called *neoteny*; when neoteny is incorporated by evolution into the normal development of individuals of a species, it is called *pedomorphosis* (Greek roots *pedo-* = 'child' + *morph-* = 'shape' or 'form of'). Pedomorphism has been immensely important in our evolution. In later chapters, we will discuss the sudden expansion of the human brain about 200,000 years ago. The shortened face, massive curiosity and ability to learn, hairlessness and soft skin that scars badly, and the ability to continue digesting milk into adult life, are all exaggerations of child-like – pedomorphic – features.

The gills of the common ancestor we share with sea squirts changed their function from filter-feeding to extracting oxygen from water, and this required a dedicated circulatory system. A muscular pumping device – a heart – drives blood forwards inside the ventral surface of the animal, then dorsally over the gills to pick up oxygen and dump carbon dioxide, then backwards to the rest of the body. This is the opposite of the arrangement found in invertebrate animals. It is not by chance that all vertebrates, including us mammals, have gills at an early stage of embryonic development.

**********

We have followed a line of evolution from a bacterium-like cell to a recognisable chordate (animal with a rudimentary backbone). Where next? I shall merely list the names of animals that lie near the stem of our ancestry and add significant features as they appeared.

The lancet, *Amphioxus*, lacks a clear head, jaws and paired fins. The fossil *Pikaia*, dates from 530 million years ago (mya) and is very like the lancet; it is the earliest known chordate, and therefore nearer our ancestor than is *Amphioxus*.

*Pteraspis* was an armoured fish that lived about 400 mya, and its armour was made of bone. Bone, which is found only in vertebrates (animals with backbones), contains much calcium, as do echinoderm skeletons – another link. The armour of *Pteraspis* took the form of bony plates in the skin, which evolved in response to attacks from a 2-metre long, scorpion-like invertebrate with ferocious pincers. Our skulls, jaws and collarbones are directly descended from those plates. Jaws evolved from the supports of gills, and small bony plates in the skin became the dense dentine of the first true teeth.

A double row of ventral fins soon reduced to two pairs, which are more efficient for paddling and steering. Then, two distinct forms of fins evolved: ray-fins, such as those found in most bony fish, and lobefins, such as those found in lungfish and coelacanths. Ray-fins look like fans, but lobefins had a bony axis with side processes.

Some of the early fish migrated to rivers and were stranded in pools during periods of drought. There is 20% by volume of oxygen in air but only 1% in water at ordinary temperature and pressure. Concentrations of animals and rising temperatures further reduced this scanty supply of oxygen in the water. Those fish that could take mouthfuls of air and absorb the oxygen from it survived better than those that could not. Those that had a large mouth cavity – a kind of sac – to hold more air did even better. Eventually, this air sac evolved into the swim bladders of bony fish and the lungs of terrestrial vertebrates.

The pools in the dwindling riverbeds diminished in size, summer after summer, millennium after millennium, so the fish had to struggle over intervening sandbars and mudflats to reach larger bodies of water. Ray-finned fish did not do as well as lobe-finned fish. We can imagine a fractional distillation in the evolution of lobe-finned fish: only the few most effective land-crawlers survived to reach larger pools in the shrinking watercourses where they lived and bred. The enormous selective pressures caused by these changing conditions led to the rapid evolution of limbs. Lobe-finned bones were reduced in number and elongated, eventually to form the fundamental pattern of four-footed animals (tetrapods): one upper limb-bone, two lower, several smaller wrist or ankle bones and five digits.

Having evolved lungs and four limbs, animals could escape fishy predators by leaving the water. But predatory tetrapods could also climb out onto a sandbank, warm up in the sun then slip back into the water to catch some cold-torpid item of food. *Ichthyostega* was an early tetrapod that probably behaved like this 370 mya. By 330 mya, vertebrates were well established on the land, which was already clothed with tree ferns and other vegetation in wetter locations, and the first insects were flying.

There then followed a vast evolutionary explosion of land tetrapods. Some remained dependent on water for their breeding and became the amphibians, such as modern newts, salamanders, frogs and toads. The rest evolved mating as a way of transferring sperms from male to female without the delicate sex cells drying up in air. A small group of these animals evolved ways of regulating their body temperature and retaining their eggs inside the body until they hatched. They evolved good sense organs, and some had whiskers, but they waddled as they walked.

Then a cataclysm struck the Earth; 250 mya vast volcanic eruptions resulted in outflows of lava about as large as Western Europe. They are called the Siberian Traps, and were possibly caused by a meteorite more than 10 km in diameter and travelling at 20 km/sec, which tore through the Earth's surface and deep into the mantle, breaking up the supercontinent that formed the land at that time, and wiping out more than 90% of living things.

As plants and animals adapted to the new conditions, there was intense competition to be the dominant land tetrapod. The internal-breeding, almost warm-blooded, waddling animals competed with egg-laying forms that got up on their hind legs, moved them under the line of their bodies, and ran efficiently. The latter became the dinosaurs, and they dominated the land, air and sea for 160 million years. Then, about 65 mya, another meteorite struck the Earth and created such a pall of water vapour and ash that winter lasted many months, if not years. The larger reptiles, or dinosaurs, could not keep warm and digest their food, and there was massive extinction all over the world.

Out of the forests emerged smaller tetrapods that had been quietly evolving warm-bloodedness, live birth rather than egg laying, and improved sense organs – the age of mammals had begun. One group of reptiles had also evolved temperature regulation and good sense organs – the birds; and they too flourished when the flying reptiles died out.

An interesting aside that throws light on the problems created by our need to classify things and the way we think about evolution is to realise that, somewhere back in our history, a single female animal laid two eggs, possibly in the same nest. One of them became the ancestor of all birds and the other the ancestor of all mammals.

A huge evolutionary radiation based on these two different body designs, mammals and birds, soon repopulated the Earth. Within a few million years there were bats and whales, cats and armadillos, horses and sloths, shrews and dogs, elephants and mice among the mammals; and hummingbirds and albatrosses, eagles and doves, ostriches and falcons, thrushes and swifts among the birds. There was also the terrifying, 3-metre-high, flightless bird *Gastornis*, which preyed on early mammals. Which of these animals became our ancestor?

Shrews.

There is an apparent sequence of evolutionary design from shrews to tree-shrews to lemurs to bush-babies to monkeys to apes to us. Of course, being alive today means that none of them could actually have been our ancestor: their body form merely indicates what such an ancestor might have looked like. Today, these animals dwell in forests where their bones quickly dissolve in the acid leaf litter. For the same reason, fossils of our ancestors are rare when they lived in forests. After we moved out onto open plains, our bodies often dried before they rotted and were quickly covered with windblown sand, and this preserved them; thus our fossil ancestors from this phase are more abundant.

**********

If we think about evidence, cause and effect, and the rules of logical argument, rather than simply and more lazily, accepting what people tell us was written in ancient texts, then this is how a plasma at 100 million degrees centigrade became a primate.

Though there are great steps in the history of the universe – the Big Bang, plasma cooling to form light elements, heavy elements forming in collapsing stars, solar systems of orbiting planets, life appearing on Earth, the modern cell, many-celled animals and the modern human mind – it is a continuum. And, what is more, it is understandable. There is grandeur in this view of life.[46]

# Chapter 5

## PRIMATES & MAN, TOOLS & BRAINS

What did our primate ancestors look like at the stages of their evolution? How did each stage behave? The first question is easier to answer than the second because bones and teeth make good fossils. However, animals are more than bones; they are also soft parts and behaviour, and behaviour in human beings is more important than structure. Understanding the behaviour of our extinct ancestors relies on three lines of enquiry.

First, how does a fossil compare with the equivalent part of a living animal? How does that living animal behave? For example, a fossil carnassial tooth (one of the pair of shearing blades in the cheeks of modern cats, dogs and related animals) allows us to assume that the owner ate flesh not plants. If the tooth was massive, flat and ridged like a horse's, the reverse was probably true. If the fossil was a long and slender hind-limb bone, we could assume that the animal was a fast runner. Other patterns of behaviour can be deduced from less direct evidence. For example, the outer layer of the human brain is folded in a lobular pattern that leaves imprints on the inside of the skull. We know that some of these lobes are associated with certain behaviours. For instance, Broca's area, a lobe of the lower left cerebral hemisphere, is involved with speech. If we found an imprint of this lobe in a fossil skull, we could reasonably deduce that the animal was either capable of speech or was evolving some precursor to it, and so on.

The second type of evidence relies on changes to the environment caused by ancestral behaviour: gnawed or broken bones, footprints, remains of nests or shelters and, most importantly for us, tools, ornaments, paintings, weapons and other artefacts.[31]

The third source of evidence for our ancestral behaviour looks at living animals that are similar to us. Chimpanzees and bonobos (which I shall collectively call 'chimps' until I need to distinguish between them) are our closest relatives by all comparisons of structure, physiology, genetics and DNA, why not also in behaviour? From fossils, it is clear that we have changed more from our common ancestor with chimps than they have.

Bringing these threads together reveals how little many of our behaviours have altered. 'Behaviour' is a general term for everything a whole animal does. Even so, the definition blurs into physiology when we consider such behaviours as respiration rate and blushing. We can conveniently, though artificially, divide an animal's set of behaviours into four parts: (1) dealing with the physical environment, such as habitat, shelter, food and predators; (2) making and using tools and other artefacts; (3) reproducing, recognising other individuals, fighting (social organisation); (4) language. Though they overlap, I shall treat each of our important ancestors under these categories.[32]

**Thirty-five million years ago**[33] *Aegyptopithecus*[47] was a monkey that lived in monsoonal rainforest, and left its teeth and bones in what is now Egypt. It was similar to the ancestor we share with modern monkeys and apes. It ate mostly fruit, and lived in groups where, as in most social animals, every

---

[31] To understand how our ancestors thought, Mithen (1996) proposed a robust model of the way this type of evidence can be used. I have drawn heavily from his work in what follows.

[32] Much of this is from Dunbar (1996), Leakey & Lewin (1992) and Mithen (1996). As a matter of style, I have written this account as if my assertions were facts. Obviously most of them are tentative hypotheses, especially deductions about a 'missing link'. Goodall (1990: p. 173).

[33] Dates of antiquity are arrived at by many different techniques, some of which are described in the works referred to.

interaction between individuals required a decision: cooperate or compete.[48]

By successfully competing with others, a male primate could improve his social standing and so increase his mating opportunities and thereby gain a reproductive advantage. By cooperating with another male, the two of them could displace a superior male and take over his access to females. The females would have to be shared between the two males; but that is evolutionarily more worthwhile than celibacy.

By similar tactics, a female could gain access to more status and food.[49] Clearly, each sex would make more useful alliances if it could remember the age, virility, aggressiveness, social position and sexual condition of other members of the group, plus who was related to whom, who were friends or temporary allies, and who were rivals. Individuals who could remember this mass of social information would be likely to leave more or healthier offspring than those who did not. Offspring inherited their parents' capacity to remember and manipulate this information, and so were also at a reproductive advantage.

Like living monkeys and apes, *Aegyptopithecus* individuals spent a great deal of time grooming each other.[50] One individual would lie down and another would carefully pick through its fur to remove lice, burs and flakes of skin.[34] Grooming feels nice because it causes pleasure hormones to be released in the brain. It is difficult for an individual to fight or betray another with whom it has recently shared the pleasure of mutual grooming. In bonobos, grooming is often replaced by copulation. In other words, grooming and sex build social relationships.

Female *Aegyptopithecus* stayed in the group, and the males moved out, just as in the monkeys of Africa and Asia today. In chimps and most human societies it is the females that disperse, leaving the males to inherit the territory. Dispersal of one or

---

[34] Seldom fleas; because flea grubs need a semi-permanent nest or den in which they can complete their life cycle. Monkeys and apes do not have such dens, though we do.

other sex increases outbreeding, and this promotes genetic diversity and vigour. Males that stay in the group, know each other from birth and, by playing together, learn who are reliable fighters, as well as their tactics. This was of great value when doing battle with other such family groups.[51]

*Aegyptopithecus* was technically inept. They could not pull leaves off a twig[52] and use it to catch termites, as modern chimps can. Like living monkeys and apes, *Aegyptopithecus* made different calls to signal to others in the group.[53] Most importantly, they used 'words' for different kinds of predator. However, having neither grammar nor abstraction, this did not constitute a true language.

Individual *Aegyptopithecus* could learn from experience, and solve problems more effectively than other contemporary animals. They used a part of their minds, called *general intelligence*, to recognise and remember important parts of their environment, such as food sources and resting places. They also had a special domain of their minds called *social intelligence* for dealing with other members of their group.[54] The domain of social intelligence handled all the highly complex interactions in the individual's daily round. *Aegyptopithecus* was intelligent and intensely curious about relationships, but relatively uninterested in its physical surroundings, other than food, predators and shelter.

It is as if *Aegyptopithecus's* mind was like a conference centre that consisted of the large hall of general intelligence.[35] Off this hall was an annex in which sat the influential subcommittee of social intelligence. Social intelligence took its own decisions and ordered patterns of behaviour without reference to general intelligence because there was no communicating door. Occasionally scraps of muffled ideas from social intelligence could be heard through the wall, but little else. For example, a male *Aegyptopithecus* would not deliberately use food to tempt a female for sex, though he may

[35] Mithen (1996: p. 72); I have modified Mithen's analogy of a cathedral with side chapels.

have learned by experience that members of their group were attracted by food, potential mates being among them.

**Twenty-three to 15 million years ago** the climate on Earth began to cool,[55] narrowing the tropical belt. In East Africa, movements of the Earth's crust cracked open the 5,000-km-long Rift Valley and heaved up mountain ranges. These mountains cast a rain shadow that caused drought, reduced the forests and opened savannah plains in that region.

*Proconsul*[56] was a monkey with some ape-like features that lived in what is now Kenya. It looked like a small chimpanzee with a long tail, and was similar to the ancestor we share with modern apes. *Proconsul* was mostly vegetarian and used no tools. Though their brains were proportionally bigger than those of *Aegyptopithecus*, their social organisation was little advanced. Their minds still consisted of the two domains of general intelligence and social intelligence, but both domains had grown larger and become more effective.

**Ten million years ago** shrinking rainforests and spreading savannah imposed great pressures on the many species of monkey and ape living in the forests. Some apes took to the forest floor and walked on the flats of their hind feet and the knuckles of the hands, as do living chimps, gorillas and older humans when they go on all fours. Baboons, other monkeys and human children on all fours walk on the palms of their hands.

**Six million years ago** there lived the Missing Link,[36] so called because it was the ancestor we share with chimps and because it has never been found. Even so, we can deduce a lot about its behaviour. It almost certainly lived in East Africa during the period of increasing aridity. Missing Link probably looked like a living chimp, and behaved in much the same way, though it was less carnivorous. It could pull leaves off a twig to catch termites, and for this it used its general intelligence. Missing Link had extended its vocabulary of calls to include more social information such as threats and invitations.

---

[36] Mithen (1996: p. 232); social attributes are deduced from living primates. See also Goodall (1990: p. 173).

Males stayed in the group and females moved out to promote outbreeding. The advantage of an integrated male fighting force increased as groups of Missing Link defended a home range against other groups and probably waged primitive warfare.[57] They shared meat,[58] but fought over a superabundance of fruit.[59] Males had a noisy display[60] in which they ran around dragging branches and throwing stones. Throwing sticks and stones gave them a great advantage when interacting with other species,[61] such as baboons and predators because throwing missiles reduced the risk of personal injury inherent in close combat.[37] Female Missing Links began to form longer-lasting relationships with particular males, copulate when they were not ovulating, and reduce the prominence of their sexual swellings when in season.[62]

Missing Link not only had all the social skills of *Aegyptopithecus* and modern monkeys, but also formed alliances to progress politically.[63] They feigned injury to curry sympathy[64] and put their young to play with those of higher-ranking parents so that they were well-connected when they grew up.[65] They sneaked out of sight to commit adultery,[66] reneged on deals, and manipulated their social images.

**********

An animal is conscious,[67] as opposed to unconscious, when it can receive sensory information. Even a single-celled animal is conscious in this sense. Missing Link was also *self-conscious*. That is to say, it was aware of what it looked like to another individual. Like chimps, if it had looked into a reflecting pool of water, and seen a leaf on its head,[68] it would have brushed it off, but *Proconsul* would not have made that connection between image and self.

Higher animals respond to different emotional states in other individuals, so they are conscious of them. More than that, an individual Missing Link could imagine what it was like to be in the same emotional state as another individual. This enabled it

---

[37] Goodall (1990: p. 209) and a striking photograph.

to predict the other's behaviour;[69] this was a major breakthrough in the evolution of minds, and we call it *other-consciousness*. Other-consciousness also included the early stages in the evolution of empathy – imagining what another individual thought of the imaginer. Awareness of its own reputation allowed an individual to adjust its behaviour to make alliances that could promote its own social position, as well as increase group cohesion. Clearly, Missing Link's social intelligence had advanced significantly. But its self- and other-consciousness were still imprisoned within the mental domain of social intelligence. Since social intelligence, self- and other-consciousness were isolated from general intelligence, neither form of consciousness could be used by an individual to improve its interactions with the natural world, or to make tools.

The beginnings of a third mental domain – *natural history intelligence* appeared at about this time. Most animals hold spatial maps of their territories, decide where to forage, observe and hunt prey animals, and identify food plants. It seems that Missing Link was able to separate this capacity from its general intelligence and social intelligence.[70] To start with, the domain of natural history intelligence was small and primitive, and hardly distinct from general intelligence.

**Four and a half million years ago**, *Australopithecus*[71] lived in eastern and southern Africa. Its preferred habitat was wooded savannah, which was becoming progressively more arid. There were several species of *Australopithecus*, some lightly built and others much larger. They could walk on their hind legs alone, though not for great distances because their feet still retained some ape-like characteristics. They ate mostly plants, and survived until one million years ago, becoming increasingly massive and vegetarian.

*Australopithecus* lived in larger social groups than Missing Link. This enabled them to better defend their territories against predators and other groups of the same species. Living in larger groups also meant that they had to know more individuals, with all the complexity that entailed. This required them to groom more individuals,[72] and doing so took up an increasing amount of time. There was clearly an advantage in grooming, for the time spent on it could have been used in feeding, watching out for predators, or some other apparently more essential activity. The advantage is that grooming builds up alliances and cements friendships, which are important in climbing the social ladder, as well as defending the group against strangers. As noted in earlier species, higher-ranking males got more mating opportunities, and higher-ranking females got better feeding.

Grooming also reduces tension in a group. For example, grooming time in chimps increases ninefold when a female is in season.[73] *Australopithecus* supplemented the physical contact of grooming with grunts and other noises to signal emotional states. They also increased their vocabulary of 'words' for different predators, as have several species of modern monkeys.

*Australopithecus* was becoming more self- and other-conscious. It is difficult to imagine the low level of consciousness that must have existed in these early stages. Maybe it was akin to a *petit mal*[74] seizure during which sufferers are still capable of carrying on normal activities, but lack any memory of having done so; they appear rather mechanical to a fully conscious observer. Even so, *Australopithecus*'s mind was very similar to that of Missing Link. Though the three domains of general, social and natural history intelligence were larger and more efficient, there was still little or no communication between them. *Australopithecus* may have used twigs or stones as tools; and simple, artificially shaped stone tools appear right at the end of their time, but we do not know who made them. It could have been either *Australopithecus* or Able Man (*Homo habilis*).

**Two and a half million years ago**, Able Man and other related species appeared in the arid savannahs of Kenya. They were 1.5 m tall and weighed about 50 kg. They had feet similar to ours and walked upright;[75] this was a profound evolutionary change. Walking upright reduces the area of back exposed to the heating effect of the sun, and allowed Able Man to see over tall grass more easily. They could carry food, materials, tools and babies that were too premature to cling on as baby apes do. Walking upright also allowed Able Man to throw things more effectively than chimps can. When compared with earlier primates, they had relatively large skulls with a brain capacity of 500–800 $cm^3$ – still smaller than ours, which averages 1,300 $cm^3$.

They scavenged carcasses killed by large cats, packs of dogs or hyenas, and maybe hunted in groups on their own account. Cats, dogs, hyenas and pigs cannot sweat through their skins; they can only lie still and pant when they get too hot. That is why most of them hunt when it is cooler at dawn, dusk or the night. Able Man avoided direct competition with these predators by scavenging or hunting when the savannah was vacant, that is in the heat of the day.

Consider: a man sets off across a hot, waterless plain in pursuit of a warthog. The warthog sprints away, but the man can still see it in the flat landscape, so keeps jogging on. In a few minutes he is sweating, and the warthog is panting. In an hour the gap has closed, and in another he comes up to the exhausted animal and kills it.

How could the man catch the much faster warthog? The warthog had been keeping cool by evaporating water from its mouth and lungs. Body fluids are more than just water; they contain salts and other materials in solution. The warthog lost too much water, and the concentration of its body fluids increased. If this concentration goes on rising beyond a certain point, blood becomes too viscous to circulate properly, the muscles and brain will be deprived of oxygen, and the animal will soon die. The warthog needed water, but none was available. If it could drink, it would be back running in a few minutes. But the man can keep going, hour after hour under the blazing sun. How can he do this? His trick is to sweat salt as well as water through his skin. The water evaporates and keeps him cool; and this is much more effective on a naked skin. The hair on his head protects his scalp from the sun, and his eyebrows stop sweat from running into his eyes. The loss of salt as he sweats prevents his blood from thickening. He replaces the salt he has lost by eating the warthog's flesh, though it takes him several hours to digest and absorb it. Now he must find water, but it is not fatal if he does not find it that day.

**********

How could a medium-sized bipedal mammal with teeth that had evolved for crushing plant food kill and butcher a large and peevish animal? The sudden transition to meat-eating could happen only if an alternative were found to meat-eating teeth, such as those possessed by true carnivores. The alternative was tools. Able Man made stone tools by striking one stone against another. This broke off a piece from the rounded stone and left a sharper angle. More blows sharpened the cutting edge further. Apart from having a sharp edge, the form of these first stone tools was determined by the shape of the original stone from which they were made and how it flaked or shattered; thus the earliest stone tools were very crude.[76] They were used for cutting hide and tendon, and smashing joints and bones.

Later, some stone tools were used to make other tools, such as sharpened sticks, which allowed Able Man to kill an animal without getting dangerously close to it. This was an advance over pulling leaves off a twig to probe a termites' nest. It required toolmakers to carry in their minds an image of the end product while doing something entirely different. Though modern chimps appear to do this, it is unlikely that *Australopithecus* could. A new, and to start with primitive domain of Able Man's mind controlled the making of these first tools: *technical intelligence.*[77] Like the subcommittees of social and natural history intelligence, there was still little communication between them, or with the large conference hall of general intelligence.

During early stages of their evolution into predators, Able Man scavenged kills from carnivores. But kills were rare, widely scattered and did not last long. It required a different kind of mind to deal with this problem. Most mammals know instinctively that the smell or sight of a predator means immediate danger, but monkeys and chimpanzees cannot deduce from seeing the track of a lion that one may be near. Able Man made this mental leap. They even realised that circling vultures indicated a carcass beneath. This involves a greatly enhanced domain of natural history intelligence.

**********

After a successful hunt, the senior Able Man hunter or another high-ranking male shared out the meat. He handed out pieces to selected individuals, or allowed them to help themselves. He chose these individuals according to sex: females in order of social standing and their offspring, male assistants in the hunt, and high-ranking males; in other words, those likely to threaten his superiority. Little has changed between this and the resident male (who paid for it) carving the Sunday joint for guests. Both are entirely different from shared feeding in a pride of lion (or *Proconsul*), which never give food to each other but treat sharing merely as tolerated theft. How much social and business negotiation is done over shared food in human society? Relative to population, probably not a lot more than by Able Man. In these simple exchanges lie the seeds of behaviour that grew into armies and trade, and would eventually make our species, *Homo sapiens*, the most widespread and, in terms of biomass, the most abundant land vertebrate ever to have evolved.

The change in diet from mainly vegetarian in *Australopithecus* to meat-eating was a profoundly important step. Brain tissue is highly active, requiring 20 times more energy than does resting muscle. More than that, the growing brain makes huge demands for food – especially fat – from the mother,[78] particularly during pregnancy and nursing. Accordingly, successful scavenging and hunting by males became very attractive to females, and so it developed into a significant form of male display.[79] The male who could bring home the largest prey item to share among the family group and associates enhanced his mating opportunities.[38] This was the driving force behind Able Man's transformation into a meat eater.

But there is another dimension. Because a male who killed the largest animals was seen by other males to be more attractive to females, his social status among other males rose.

---

[38] This persists today. Adultery seems to be epidemic among fox hunters (personal observation).

So males competed with each other and hunting became a form of display long after the food requirements of the family were met; it seems as if the hunt evolved into a masculine way of showing off to enhance social status, like the evolution of stags' antlers. Maybe that is why they are hung on the walls of modern hunters' homes.

Men hunted large mammals and women gathered fruits and vegetables, which are equally nutritious for adults. In terms of energy expended against energy gained, gathering is more cost-effective than hunting large mammals. However, plant material contains less fat so it is not as good for children's growing brains, and less prized for this reason. A prey animal takes evasive action whereas plants do not; thus, if a hunter failed, he was regarded as unlucky;[80] but if a gatherer failed, males treated her as if she were merely lazy. With all this emphasis on fat, why was it important?

The human brain is composed almost entirely of nerve cells. Nerve cells consist of a cell body that is grey in colour, and a long, white extension that carries nerve impulses to other nerve cells. Nerve cell bodies are mostly in the outer layer of the brain, giving it a grey colour and the name 'grey matter'. The impulse-conducting extensions of nerve cells are mostly in the inner parts of the brain. To stop impulses jumping from one extension to another, they are insulated by fat. This is why there is so much fat in brain tissue, and a lot of fat is needed in the diet of a child whose brain is growing fast.

Our brains are divided into three regions: forebrain, midbrain and hindbrain. It was not the whole of Able Man's brain that became larger, only the grey matter of the forebrain. I shall call this the 'grey brain'. What do we share with other species of mammal that also have large grey brains, when compared with their relatives? Bizarrely, vampire bats[81] are a good example. They live in colonies and feed at night on blood lapped from other animals. Sometimes they are unlucky, and cannot find food one night. When a hungry vampire gets back to the roost, it asks for food from members of its colony who have been more successful. If it has shared its good fortune in the past, it is more likely to be fed than if it had refused to help others. Clearly vampires need to remember not only who is who in the colony, but also who has been generous and who miserly. To do this they need big grey brains; compared with other bats, they have.

Larger animals need larger brains, so a simple comparison of absolute brain size is not very helpful. More useful is the ratio between grey brain and whole brain. Vampires and humans have a high grey brain to whole brain ratio and are social, but concurrence does not prove relationship. Since no other bats have big grey brains, our common ancestor with vampires was unlikely to have had a big grey brain, so it looks as if big grey brains have evolved more than once. Big grey brains also occur more frequently in primates with whom we share a more recent ancestor. This suggests that our common ancestor – Missing Link – also had a grey brain that was large when compared with that of near relatives.

Chimps are very like us in that they have friends – individuals they sit near to, share food with, whose hair they groom, and whom they can rely on for help in a fight. They also remember who is friendly with whom, and who their rivals are. It all gets very complicated in a tribe of about 100 individuals. This is why we have such a huge brain: to remember not only each person's track record, but also the intricate web of family and social relationships.[82]

How social are other primates? Is there a connection between some measure of their sociality and the ratio of their grey brain to whole brain? Yes there is. All monkeys and apes except for the orangutan live in groups. The bigger the social group in a species, the higher the ratio of grey brain to whole brain. Indeed, this measure is accurate enough for us to predict that the average group size of *Australopithecus* was 67 and Able Man 82. The figure for modern chimps is 60, and 150 for us.[83] Individual variation seldom exceeds 50% of each number, either way.

One hundred and fifty individuals in the basic human tribe? How many people do you send Christmas cards to (including people at one address)? How many people do you know socially? How many names in your personal address book? I bet that each of these numbers is more than 10 and less than 1000. I will go further and predict that it is between 50 and 200. That is about as many as the social parts of our brains can handle. In September 2015, the mean number of 'friends' of the 1.5 billion subscribers[84] to the social networking website *Facebook* was 338,[39] and the median was 200.[85] These are within the 50% more or less bracket.

These numbers do not mean that our ancestors habitually went around in groups this size; they probably lived most of their lives in smaller family parties of 4 to 10. But there was a great advantage in knowing the individuals in a bigger group when it came to finding and sharing a large carcass, which one family could not hope to eat before it decayed. Having lots of friends was also useful for defending the carcass against hyenas, jackals and rival bands of people. Larger groups of individuals cooperating together could defend territories against attackers more successfully than could smaller groups. Effective cooperation means knowing who is who, who owes what to whom and a mass of other social facts. Large brains are essential for remembering and processing all this information. The earlier a child can begin learning social skills, the more time it will have to practise using them; therefore it is an advantage to have a big brain early in life, and preferably being born with one.

In four-legged animals, the gut and womb fall forwards and are supported by the rib cage. The hind-limbs are attached to the spine by the pelvis, which, in this body posture, can have a large rearward opening. When the spine is vertical, the gut falls downwards into the pelvic basin, and imposes a heavy outward strain on the body wall, and this makes human beings prone to pot bellies and men to hernias.[40] We partly solved this problem by evolving a basin-shaped pelvis with a narrow opening between the legs.[86] However, there is such an enormous

[39] In summer 2013 the figure was 120.

[40] A hernia is caused by the intestine bulging through a weakening in the body wall where the testes have descended into the scrotum through the inguinal canal.

advantage in learning early on in life that babies and children have disproportionately large brains and skulls that must pass through the narrowed pelvic opening during birth. This is why human birth is such a long and difficult process compared with that in other mammals of similar body size.

Human babies being born at an earlier stage of development reduces the risk of birth difficulties, but means that they are born in a very helpless condition. Compared with us, baby chimps have small skulls and are capable of seeing, vocalising and running about in a few days. They are born at a stage that is only reached by human babies a year and a half after birth. Female chimps are far less disabled by pregnancy than we are, so can go on feeding and keeping up with the troop. The troop is mutually supportive – certainly among females that are friends.

Chimps do not have monogamous relationships – most adults in a group copulate promiscuously.[41] This means that every male in a group could be the father of any infant. Therefore, it is in the genetic interest of every male to act as if all babies were his own, and so to care for the group as a whole. It is interesting to note that gibbons – our most closely related minor ape – are generally monogamous and adultery is virtually unknown.[87]

Large baby skulls, difficult birth and a prolonged period of infant dependency mean that the human mother cannot gather or hunt her own food in the later stages of pregnancy, nor for a while after the baby is born. Yet she needs meat and fat to feed her baby's demanding brain. The only way the baby and she can survive is to depend on help from others. Either the group – effectively a family unit – operates as a whole, or each man looks after what he thinks is his own baby. Clearly we experimented with both reproductive strategies, promiscuity and

[41] This behaviour refers specifically to bonobos (pigmy chimpanzees). In chimpanzees, it is the dominant male that has most mating rights, though females will mate with any male, and males will mate with any female, if they get the chance.

monogamy, during our evolution, and are settling on predominantly the second; this led to pair-bonding.

By having just one female bonded to him, a male did not have to spend so much of his time competing with other males for mating rights, as chimpanzees do. This allowed him to cooperate with other males for hunting and defending the family group. Associated with the evolution of male cooperation and pair-bonding was the appearance of silent oestrus.[88] Since mating takes up feeding time, and distracts from watching out for predators, there is survival value in confining it to the time when there is an egg ready to be fertilised. This is why most female mammals signal that they are in season by scents, visual signals and behaviour. Silent oestrus is when these signs are suppressed, and the female is willing to mate at practically any time. Mating is pleasurable, so rewards the male who cannot tell whether his female is ovulating or not. Just think of the chaos in a crowded city if women would only have sex when they were ovulating; and, when they were, their faces turned bright pink.[89] Men would have to escort their ovulating wives everywhere, and non-ovulating women would want to keep a very sharp eye on their husbands. Cooperation between men would be sorely tried; indeed, cities would never have evolved. Even with pair-bonding and silent oestrus, both men and women indulge in extramarital sex. Indeed, it has been claimed that about 4% of children in one recent study are not their legal father's.[90] 'Mummy's baby, daddy's maybe.' How could a man, who had been away hunting, be sure that the woman he was caring for was carrying his child? What better way than to listen to gossip? But gossip requires subjects, verbs, objects and tenses – language.

# Chapter 6

## FIRE & LANGUAGE, FARMING & WAR

**Approximately 1.8 million years ago**, the tropical belt of Africa narrowed then broadened as the climate in Asia and Europe switched between icy tundra and thick temperate forest.

These cold periods were not what are popularly known as the Ice Ages; they came much later. Upright Man (*Homo erectus*) evolved in Africa and was the first of our ancestors to leave it. Bolder tribes of this species crossed Sinai, which was not a desert then, and spread out into Europe, where they survived until 300,000 years ago. Others reached China and Java, while the original stocks remained in East and South Africa.

Upright Man was taller and more robust than Able Man, and had a brain capacity of 750 to 1250 $cm^3$. The striking thing about their appearance was that, even though brain capacity had increased, their features were much more ape-like than ours, with low forehead, prominent eyebrow ridges and no chin. On the other hand, below the neck their bodies were almost identical to ours.

Our bodies have evolved into a form that is not particularly good at doing anything well. Monkeys can climb more skilfully, kangaroos can jump higher and further, most animals of our body size can run faster, and dolphins swim more swiftly. But imagine an obstacle race with all types of vertebrate animal lined up at the start. Suppose that they all have to run 10 km, swim 1 km, climb over a smooth wall 3 m high, jump across a chasm 3 m wide, climb a tree, climb up a 10-m-long rope, climb the face of a sand dune, cross a boulder field and crawl under a tarpaulin. Only one body form of all the world's vertebrates would finish the course: ours. We are the supremely adaptable animal.

Upright Man began to use fire.[91] This was a huge leap forward in behaviour, and it immediately separated our ancestors from all other species that had ever lived: *we use more energy than we eat*. It let us control forces far greater than our own bodies, and we used them to modify our environment quickly. Such a technical skill gave us a huge advantage over all our competitors. It also enabled us to live in caves. Previously, predatory cats and bears could keep us out, or corner us in a cave, and we had no defence against them. As soon as we overcame our fear of fire and could handle it, we dominated them and were instantly on the way to becoming top predator.

With increasing aridity in East Africa,[92] the vegetation became more flammable. It is easy to get rid of big cats and other predators by setting fire to grassland, and at the same time drive large grazing animals into ambushes of men with sharpened sticks. Maybe our taste for cooked meat began that way. Cooking certainly widened the range of foods we could exploit. Setting fire to the countryside also began the vast impact we have had on this planet, and ultimately what this book is about.

**********

Upright Man made pear-shaped hand-axes,[93] which were a great advance over the crude tools of Able Man. An astonishing feature of these hand-axes is that they are all similar. It is astonishing when we remember that the stones from which they were made varied widely in shape and texture; that the people who made them ranged from South Africa to Java via Europe and China; and also that the design hardly changed in a million years. With our modern mental furniture, we are bound to ask why the hand-axes did not evolve as our modern tools do.

This unchanging design throws light on how Upright Man's mind worked. In making a hand-axe, the makers could imagine the shape of the finished tool as they worked. They adjusted each blow according to three parameters: the image in their minds; the texture of the stone they were working on; and the stage in the process they had reached.[94] Also relevant to our understanding is that Upright Man may have used wooden tools, such as sharpened sticks for spears, though no such tool has survived, but they did not use compound tools, such as hafted axes. From this information, it seems that they had greatly extended Able Man's domain of technical intelligence. However, as far as we know, they did not use bone, horn or ivory, which are animal products. Therefore, Upright Man thought about these materials in his domain of natural history intelligence, not in the domains of technical or general intelligence. This signals clearly that these three mental domains were still isolated from one another.

**********

As we have seen, sharing the pleasure of grooming and being groomed were essential for maintaining social structure in all higher primate groups. When a group got too big for this cohesive balm, it tended to break up. Good hunting territory was fixed in area; and, when it was all occupied, splinter groups had to compete for the use of it. Given roughly equal weapons, larger groups were more likely to win in fights with smaller ones. Therefore, there was a strong advantage in belonging to a large group. However, large groups tended to break up because of insufficient grooming time.

Tensions build in larger groups where time needed to groom cuts into feeding, resting or keeping a look-out for predators. The problem is that grooming can be done only on a one-to-one basis. Though individuals can only physically groom one other individual at a time, they can make grunts to a third while doing so. Indeed noises, if loud enough, can reach quite a crowd. But the noises must convey more than just 'Look at me'. They need to relate to something already within the hearer, something which, when stimulated, makes them feel as if they were being groomed. Other-consciousness is an essential pre-requisite for this. Such a series of sounds evolved in the domain of social intelligence and eventually became language.[95] Language is such an important feature of human beings that it is worth digressing to look at it here, and take it beyond what Upright Man could achieve.

First, what is human verbal language? How is it different from other forms of communication in animals? Animals communicate by touch, scent, visual signs and sounds. We are concerned here only with sounds and their perception by ears. (Writing that uses hands and sight came much later.) Many animals make noises that other animals respond to. The hiss of a snake warns an aggressor. Bird song attracts mates, while bird calls warn of predators. These sounds express emotion in the animal making them; they do not refer to a particular object or event outside that animal's body. Some species of monkey have distinct calls for several different predators: eagle, leopard, snake, and so on.[96] Even though these sounds are still expressed as a result of an emotion, such as fear, they do refer to specific things outside the caller's body. We modern humans have some difficulty in understanding this because, for us, words convey a mental representation that is not present in monkeys; nor was it in Upright Man at the dawn of language.[97] Our difficulty is due to the sounds they make beginning as nervous impulses in the hindbrain, whereas our speech is controlled by our grey brains, which are a part of the forebrain.

The vastly larger and more complex grey brain, allows for grammatical structure.[98] Two areas of the grey brain have been identified as concerned with language: Broca's and Wernicke's areas.[99] If my Broca's area were damaged, I would have severe difficulty with grammar. If my Wernicke's area were damaged, my grammar would be correct, but I would talk nonsense. Together, these two areas are concerned with the formation of a concept, remembering a word associated with it, storing verbs – regular verbs separately from irregular verbs – and storing nouns. It looks as if we are born with a capacity for language already in place. We are also born with the ability to add to that capacity by learning.

Monkeys also have Broca's and Wernicke's areas in their grey brains, but they use neither for speech. Instead, the areas discriminate sound sequences and control muscles of the face, mouth, tongue and larynx. As the grey brain grows, its folding leaves impressions inside the skull bones. Rudimentary lobes, which correspond to those of Broca's and Wernicke's areas, can be seen in Upright Man's skulls. Their form suggests that Upright Man's sound communication was advanced over that of monkeys and chimps, but it still lacked grammar. It is more probable that, like monkeys, these areas were used for other, though related, purposes that were subsequently adapted for language.

Since the *capacity* for language (not the language itself) is innate, it must be genetically controlled.[100] Two lines of argument follow from this idea. The first is that, since it is genetically controlled, the capacity for language must exist in the form of permanently arranged nerve connections; however, these are not the only hardwired circuits in our brains. The words, grammar and syntax of whatever language we hear in our early years are fixed in new and permanent nerve connections,[101] but these permanent connections will not be passed on to our children genetically because they were acquired during our individual lifetimes.

The second line argues that, since the capacity for language is genetically controlled, it must have evolved. Since it has evolved, it must have given our ancestors some reproductive advantage. This advantage is likely to have been that, the sooner in their lives our ancestors learned their language, the more sophisticated their use of it could become. The more brilliantly and wittily they spoke,[102] the more attractive they were to potential mates and the greater were their mating opportunities. Therefore, selection favoured there being plenty of genetically controlled permanent nerve connections in place before the learning process added more. That is *how* the capacity for language may have evolved. It looks as if learning has guided natural selection;[103] but then, other behaviours often do. The next question is: *Why* did language evolve?

The first clue comes from asking what Upright Man talked about. Coordination in a hunt? No: other social predators, such as hunting dogs and lions, manage perfectly well without talking about it. The real answer is gossip.[104] They used language to bond together in groups of individuals whose reputations they knew and could rely on, and to find out about other people. Social sniffing was vitally important: 'Do you know so-and-so? Oh really? Jolly good sort, y'know.' We assess our relative social positions, and find out who is on whose side by small talk and gossip. We also discover who knows what, which is the ultimate basis of power and influence.[42] An important, perhaps *the* most important, function of language was to put out propaganda about oneself and other people;[105] that is, to manipulate our own and other people's reputations as modern politicians and media persons know so well.

Language is also a very good way for a man to find out whether his woman had been faithful while he had been away hunting or working,[106] then to make a judgement on the probability of her child being his: hence the socially vital importance of a woman's reputation. Likewise, language is equally important to a woman to find out whether her husband had been wasting his time feeding another woman and child. Supporting evidence for this is that men are furious when their wives have sex with another man, but do not mind some emotional entanglement; whereas women do not mind their husbands having a casual affair, but are very unhappy about emotional commitment to another woman.[107] Telling lies seems to have been an early and essential function of language.

Detecting other cheats was another driving force in the evolution of language.[108] People sharing a tribal language are usually related to one another, and they cooperate by doing each other favours. Gossip is a powerful means of policing cooperation and conformity to society's norms. An outsider is more likely to take without receiving, and might even be a spy from another tribe with invasion in mind. Tiny inflections of accent would reveal his origins. A common English password during World War II was 'squirrel', which central Europeans find difficult to pronounce in the English way. How many public faces have been betrayed by the shape of their vowels – to those who think that they matter – particularly if they have evidently tried to change them. Today, in the early decades of the 21st Century, it is amusing to notice the slight embarrassment of politicians and other public figures when they forget, in spite of obviously careful rehearsal, to convert their natural 't's into glottal stops. I was amused by the evident embarrassment on the

[42] John Chapple, personal communication, November 1994.

face of Samantha Cameron, the Prime Minister's wife, when, in a staged TV appearance before the British 2015 General Election, she said, 'Me and the children ....' I suspect that public relations advisers had told her that the grammatically more correct 'The children and I ...' would lose votes for her husband. Who else thinks they need to give the impression that they are one of the people? 'Aping' the powerful is a precise biological concept; the mass of people in a democracy being more powerful than a prime minister.

Language got a powerful boost when employed directly in sex. Smiling, jokes, wit, music, poems and words of endearment are all releasers of pleasure hormones in the brain.[109] Therefore, displays of linguistic skill are regarded as sexy as well as uniting a group that enjoys them. Words of songs, such as those used by clubs, teams, religious sects and war parties, also strengthen group unity.[110] The *haka* of the New Zealand rugby football team is a good example. War songs without music seem to be the content of much parliamentary debate, where agreement seems scarcely tolerable.[111] In sum, language evolved for social reasons; only later on it was used to talk about technical problems and natural history. Small wonder that it is such a poor medium for describing the world outside ourselves,[112] but it is the best we have – apart from mathematics.

Even though our capacity for language is hardwired at birth, we must still learn the language itself. Each of us learns the melody, stress and timing of our mother's language while we are still in her womb.[113] At birth, our voice box is high in the throat, which means that there is no room to move the tongue backwards to make vowel sounds – only baby cries. By three months the voice box has moved down to its adult human position.[43] Even then it rises to close off the windpipe when we swallow.

[43] Pinker (1994: pp. 238, 286). This is one highly significant result of walking fully upright. The larynx moved downwards in the throat and became a

Six months after birth we begin to sort and combine the different sounds that form the units of our native language. During our first year we play with the sounds we hear, string them into repetitive burbles and babbles, and discover how to move our tongue, lips and vocal cords to make the right sounds. By 18 months of age, we utter our first words,[114] and soon we are using about 50. By two years, we have a good vocabulary, but nearly all our words are nouns with only a few verbs. In the next few months we start putting nouns and verbs together, and words in their right order. Only later do we add prepositions, pronouns and other parts of speech. By three years, we are chatting away in simple sentences and using 1,000–3,000 words. By six, we have a vocabulary of 13,000 words and have mastered the grammar we are exposed to. By 18, we have 60,000[44] words and are equipped to discuss the most abstract concepts. We have opened the door to ideas. However, more of all this later, because Upright Man was only just beginning to talk about social matters.

**Six hundred thousand years ago**, Heidelberg Man or *Homo heidelbergensis* appeared in Europe and survived there until 100 thousand years ago. They were large and robust with a brain capacity of 1100 to 1400 $cm^3$. Their eyebrow ridges were smaller than in Upright Man from whom they were descended. Very few remains have been found, so little is known of their behaviour.

**One hundred and fifty thousand years ago**, Neanderthal Man *Homo neanderthalensis* evolved from Heidelberg Man and survived until 30 thousand years ago, which was almost to the end of the Ice Ages. The remains of Neanderthal Man are abundant; they are found only in Europe and West Asia. Their brain capacity was 1200 to 1750 $cm^3$ and their eyebrow ridges were smaller still than those of Heidelberg Man. They had short legs and massive arms and chests, and suffered many injuries

---

voice box. This new position allowed, but did not cause, the ability to make vowel sounds.

[44] Dunbar (1996: p. 3), but Encyclopaedia Britannica (1991: vol. 9, p. 682) gives 2,500 words at six years old.

through a physically demanding and predominantly hunting way of life.

They lived on tundra, hunting and scavenging reindeer, red deer, wild cattle and horses. They used fire, and probably invented then exploited the new ecological niche of thawing frozen carcasses. From the poverty of modern tundra, it might be assumed that food and fuel resources were rare and that Neanderthals needed vast hunting areas; however, the sheer density of mammoth bones in Siberia suggests that the tundra habitat once supported a flourishing large mammal fauna. You will see in Chapter 10 on ecology, that minerals cycling within an ecosystem with little export can become rich; it is only when scarce elements, such as phosphorus, are removed that ecosystems become impoverished, causing the eventual extinction of many large herbivorous mammals, such as mammoths and woolly rhinoceros.

Neanderthal Man had a gestation of nearly a year,[115] but our inflated skulls at birth made it essential that we were born at an earlier stage of development. This did not mean that our newborn were relatively smaller; at birth, both our and chimps' babies are about 5% of their mother's body weight. Being born earlier in development means that the bones of the skull can be deformed more easily and so pass through the birth canal in the pelvis without damage; this required that our gestation be reduced to 10 lunar months. In effect, we need a year and a half of growth outside the womb to reach the developmental stage of a newborn chimp. This meant that our mothers had to produce more milk for longer,[116] so they needed a diet richer in protein and fat.[117] The bald fact is that women traded meat for sex with their pair-bonded men.

Neanderthals lived in larger groups than did earlier human species and this suggests that their language was more advanced. They inhabited caves and rock shelters, and wore skins, but could not sew,[118] nor did they use bone, antler or ivory for tools. The stone tools Neanderthals made represented a quantum leap in technique. They not only improved on the traditional hand-axe by using different materials to strike flakes off a core, but they prepared suitable stones so that a single blow would yield a ready-formed spear point or blade. Even after years of practice, few modern flint-knappers have been able to emulate Neanderthal skills in forming symmetrical, effective, leaf shaped arrowheads that were also simply beautiful. Neanderthals had a deep understanding of the physical properties of their working materials and a clear mental image of what they were trying to achieve in toolmaking. For example, in making a spear point, they obviously had a spear in mind. A spear is a tool consisting of two parts: a point and a wooden shaft; thinking of any composite tool is in itself a major mental advance. There is no evidence that particular Neanderthals became specialists, though some individuals probably had greater skills than others. It is also likely that individuals exchanged the tools they had made for other goods.

The mental domains of natural history intelligence, technical intelligence and social intelligence had all expanded and begun to overlap more with general intelligence. Neanderthals could think very simply about technical and environmental matters, but could not talk about them because language remained devoted to social matters, as it still does in many modern people.

Neanderthals hung on in Europe until 30,000 years ago, then disappeared in the face of invading Modern Man (*Homo sapiens*). Neanderthals seem to have been an evolutionary side branch of the human family. No hybrid skeletons have been found,[45] which suggests that we did not regularly interbreed with them, even though they and we coexisted for several thousand years. Recent DNA studies seek to decide this.[119]

[45] Diamond (1991: p. 44). Even though there is some recent evidence that *Homo sapiens* interbred locally with *Homo neanderthalensis*, I have treated the two as separate species, not as the subspecies *Homo sapiens sapiens* and *Homo sapiens neanderthalensis* as used by other authors. See also Leakey & Lewin (1992: p. 232). Geneticist Peter Forster from Cambridge University and mathematician Arne Röhl (University of Hamburg) presented a phylogenetic network, published in *Molecular Biology and Evolution* in 2001, with evidence for mixing of *H. sapiens* and *H. neanderthalensis* genes. Alan Templeton (2002), *Nature*, analysed 11 genetic trees, obviously on the basis of mitochondrial DNA, and also found local interbreeding (personal communication, Wolfgang Hickel, 2002).

Soon after Modern Man appeared in Europe, some Neanderthals seem to have copied their tools briefly before dying out.[120]

**Two hundred thousand years ago**, Modern Man (*Homo sapiens*) appeared in southern Africa. Our[46] skeletons were less robust than Neanderthal Man's; eyebrow ridges were almost absent, a chin had evolved and brain capacity increased to 1200 to 1700 $cm^3$. Skeletons from Ethiopia have been dated to 195 thousand years ago,[121] which indicates that we had already spread throughout Africa.

The next 70,000 years were a time of supreme trial for our species because the African climate cooled and dried causing deserts to spread. Our population fell to a few hundred individuals that dwelt in coastal caves in South Africa and foraged along the shore.[122] Being so reduced in numbers diminished our genetic diversity; a phenomenon called 'an evolutionary bottleneck'. Because there were so few alternative genes for selection to act on and the mutation rate was so slow, our bodies could not evolve genetically to cope with our changing environment. As the imaginary obstacle race described earlier reveals, our bodies were well adapted to cope with almost any physical challenge, even though not brilliantly. Our huge advantage over other animals lay in our ability to adapt behaviourally, and this allowed us to expand throughout Africa. During this period of expansion, our tools and behaviour suddenly diversified. We made a wide range of stone points, flakes, blades, arrowheads and axes, needles and beads. More than that, we used bone, horn and ivory as materials, and we placed artefacts, such as beads, and parts of animals in our burial places.

It seems that all five domains of the mind – general intelligence, natural history intelligence, technical intelligence, social intelligence and language – had finally combined.[123] The subcommittees had left their private rooms and entered the conference hall for a plenary session. Now we could understand

[46] In this chapter and for the rest of the book, I use the first person plural to mean modern human beings.

that our habitat, our food, our tools and our relations with our fellow human beings all interacted, and we could think about them together. We could set a trap for an animal, and seduce with a present. A trigger for this explosive increase in adaptive behaviour was probably our beginning to consciously teach skills to younger people.[47] [124] In this way, our cultural heritage took a huge forward stride.

From this point on, the evolutionary pathways of our genetic and cultural evolution diverged and can be thought of separately. As we saw in Chapter 4, our genetic evolution takes place over millennia by individuals breeding, varying and being selected. Very gradually, the genetic composition of a population of individuals alters. In sharp contrast, culture is learned behaviour that is passed on from one individual to another; and that can happen in seconds. Notice that it is our *behaviour* that is evolving and *not* our genes. As soon as we appreciate that a behaviour can multiply and vary and be selected just as a gene can, then it is clear that Darwinian evolution is at work. As individuals, we cannot change the genes each of us carries, but we can instantly change our behaviour, and by our own free will too. This is what Richard Dawkins means by 'We, alone on earth, can rebel against the tyranny of the selfish replicators'.[125] Herein lies our future as a species.

**********

We dispersed throughout Africa and, as our population pressures increased, we adapted to life in desert and rainforest, savannah and tundra, seashore and mountains. There was still sufficient genetic diversity for our bodies to evolve into both 130 cm-tall pigmies and 220 cm-tall Sudanese. Where the sun was hottest, we retained melanin, the black pigment in our skins, to absorb ultraviolet radiation and prevent skin melanomas. Melanin synthesis releases vitamin D, which is essential for calcium and phosphate metabolism. Those of us

[47] Young chimps copy the behaviour of older individuals who do not *teach.*

who migrated north to darker climates lost the permanent deposit of melanin in our skins and supplemented our dietary vitamin D by making its production facultative in sunlight.

A few adventurous tribes crossed Sinai into Asia Minor. This was a momentous journey for our species; momentous because so few people made it – maybe only a few thousand. Once again we went through another evolutionary bottleneck, which reduced our genetic diversity even more. Some tribes turned eastwards and returned to Africa, possibly over the Strait of Hormuz, bringing Eurasian cattle with them. They moved up the Nile and fanned out over the eastern half of Africa. Other tribes that left Africa spread into Asia, and encountered an extraordinary phenomenon: the large mammals and flightless birds we met did not recognise us as predators.[126] They had evolved over millions of years being preyed upon by conventional solitary or pack carnivores. They did not learn until it was too late that a bipedal animal without obvious teeth or claws could be such an effective predator.

It is a salutary experience to be watching herds of game grazing or cudding peacefully on an African plain when the vertical figure of a man appears like an admonishing finger. In an instant the herds are alert and on their feet, if not thundering away. Contrast this response to tourists walking among tortoises and nesting birds on the isolated Galápagos Islands. These island animals are oblivious to any possible threat. Indeed, a flycatcher treated the tallest person in our group as a convenient lookout perch. They have had only 500 years of killing by people; hardly time for fear to become genetically ingrained. The animals we preyed on could not cope with groups of intelligently cooperating men that used fire, spears, missiles and traps; indeed, they behaved like modern domestic animals. With so much easily obtained meat on the hoof, our population grew and we spread rapidly, reaching Australia and Europe 40,000 years ago, the Americas 15,000 years ago, and New Zealand 1,000 years ago.[127]

As we arrived in each of these continents, we hunted the large animals that were easy game,[128] and switched to smaller species when the large ones became rare. These alternative food sources allowed us to keep on pursuing the large animals until the last one was gone. This was the beginning of the sixth great extinction to befall the wildlife of the world.[129] All other extinctions had been caused by abiotic cataclysms, such as meteorites striking the Earth, perhaps with volcanic activity that caused sudden climatic changes. The global extinction that we began (and are continuing with increasing effectiveness) is the first to have been caused by a single living species. An example of our wanton destructiveness is to be seen in the skeletons of tens of thousands of wild horses found at the base of a cliff at Solutré in France,[130] evidently trapped and slaughtered by our ancestors 25,000 years ago. The harvest ends abruptly. The great flightless birds of Madagascar and New Zealand[131] all became extinct within a few generations of our arrival. Fifty-seven species[132] of large North American mammals became extinct shortly after we crossed from Asia to Alaska. That is why there are so few really big mammals in the Americas. The list goes on, except in Africa because that is where we and our prey animals evolved together.

**********

Group size in modern human societies follows a clear pattern, whatever the race or culture. Here are some figures,[48] which can vary by 50%, together with some examples:

- 4: small family unit; one conversation at a time in informal situations; seats in a motor car; doubles court-game;
- 10: circle of intimate friends; section of a military platoon; jury; inner cabinet of government; sports team; Apostles; seats in a minibus;
- 150: living descendants of one couple plus in-laws in a poor country; clans that associate for ritual functions; people a person knows and can name; average number of Christmas cards sent; addresses in a personal address book; average number of friends on *Facebook*; military Company, including Roman

[48] Dunbar (1996: p. 76); '150' is called 'the Dunbar number'.

*centuria*; people in West Asian Neolithic village; people in modern horticultural villages in Indonesia, the Philippines and South America; average Hutterite community; personnel in a modern business run on informal lines (formal management is necessary above 150);

- 1500–2000: tribe; crowd one person can address by shouting; theatre audience; clients in a salesman's business list.

These group sizes were beginning to be in place by the end of Upright Man's time.[49]

We lived in closely knit family groups of 4–10 individuals, and knew by sight about 150 other people whom we met from time to time in our wanderings. Our own family was often related to the others we met, so we shared skills, traditions, attitudes and assumptions. 'So' needs some explanation, or at least a reminder: the evolutionary history of favouring relatives before strangers goes back to the beginning of life. Even though our evolution was becoming increasingly behavioural, strong genetic ties remain and they persist today – as nepotism.

Disease was not a problem because we lived in scattered communities and kept moving, thus giving infections little chance to spread. The life of hunting and gathering was not easy once we had eliminated the larger mammals and birds. We had as many children as we could, and the weak ones died if there was not enough food, or they were injured or had some genetic defect. We travelled around the territory we occupied, using up the resources we needed. These resources regenerated when we moved on. Thus the environment imposed limits to the density of people.

Clearly, the resources we required could not regenerate if other groups came into the area as soon as we left it. If we found other families in our territory we warned them off. If they

[49] As an interesting aside, I have noticed that the smaller the group of people speakers are addressing, the more sense they make.

resisted, we fought them even though fighting ran the risk of death or injury. Those families that evolved displays of strength that intimidated other families and made them retreat before fighting, suffered fewer injuries and so were likely to leave more offspring. This also suited the vanquished because they could regroup more quickly after retreating than if they had suffered losses in battle. This was not an easy option when the alternative was starvation.

Even though our ancestors soon evolved patterns of behaviour that reduced conflict, some was unavoidable. Fighting is ritualised in many species of animals armed with lethal teeth, claws or horns, and death or serious injury is often avoided by signals of submission and their acceptance by the victor. However, we fought other family bands by throwing spears and projecting arrows from bows or stones from slings. These weapons made contact with an opponent at a distance and so reduced the risk of injury to the thrower. It also meant that the thrower did not perceive signs of submission before the attack, thus we tended to fight to the death. However, in battles between rival bands of young male chimps, they would clearly kill each other if they could, and only lack of weapons prevented them.[133] So maybe the idea of killing is more deeply engrained in us, and not just the product of making contact at a distance.[134]

Our ancestors had evolved language so that they could groom more than one another at a time, and this had allowed them to increase the size of their family group while maintaining social bonds. This behavioural device was overtaken by another more crucial restraint: the dispersed nature of our food forced our populations to remain scattered. Thus, assembling a force when we were attacked inevitably took time, which could be too late. Given that weapons were much the same among all tribes and families, for example, all had bows and arrows, the outcome of conflict depended on the number of fighters on each side.[135] It took little imagination to see that herding the animals we hunted and protecting the plants whose fruits we harvested would overcome the food problem and allow higher density of population. So began farming.

Farming began independently in five or six centres throughout the world: West Africa, Central America, East Asia, New Guinea, one or more Pacific islands and the Near East.[50] The archaeology of the Near East is best known and the story of how it came about throws light on our present predicament.

Eleven thousand years ago there was a vast seasonal migration of gazelles through what are now Syria, western Iraq and Jordan.[136] Judging by bones found in middens, gazelles formed 80% of the meat eaten by people living in the area. By 8,500 years ago, they had learned to trap the animals in stone-walled pens, and the proportion of their meat in the human diet had fallen to 20%. Before much longer, gazelles had become so scarce that it was not worthwhile hunting them. The human population had increased so much that people there had to find other sources of food.

Hunters established a territory in which wild animals were regarded as 'belonging', so the hunters protected 'their' game from predators and other tribes of people. A hunter 'owns' an animal he has just killed, and reserves the right – under the conventions of his tribe, which were derived from our non-human primate ancestors – to share it out among his family and friends.

At the hunter-gatherer stage of our evolution, we did not recognise ownership of living plants and animals. We thought that, until a plant was gathered or a large animal killed, it belonged to no one. The idea of looking after a living prey animal by protecting it from predators and competitors, and ensuring that it had shelter, food and water, seemed a pointless expenditure of energy when game was abundant, but understandable when it was scarce.

Gazelles are practically impossible to manage in the open because they scatter when threatened. It is much easier to round up large animals that bunch together when alarmed. Cattle,

---

[50] Specifically: Syria, Lebanon, Jordan, Israel and Iraq with parts of Turkey, Iran and Saudi Arabia. See Clutton-Brock, Juliet (1999: p. 67) and Ucko & Dimbleby (1971).

sheep, goats, lamas, buffalo and horses all do this, so are suitable for domestication. Managing animals required an investment of time and effort; and to protect that investment, we evolved a sense of ownership that was distinct from territoriality. Likewise, why tend plants when there were plenty to be gathered?[137] Cultivating plants requires an even deeper sense of ownership than do animals because they cannot be moved to places of greater security, as can a herd.

Our crops were seasonal, so there were long periods between harvests; this meant that we had to store enough food to tide us over the lean part of the year. We were also tempted to produce and hoard more than that as security against crop failure.[138] In addition, a well-filled barn became a way of impressing women looking for a mate.

Storing food was an advantage in that it allowed a still greater density of population, but it also became a resource that could be stolen. If our village crops failed or stores were destroyed by insects, rodents or water, we were faced with the choice between starvation and asking a neighbouring village to share their stores.[139] If they refused, our choice was simply altered: starve or take the food we needed by force; however, force would only succeed if we could muster enough able-bodied people.[140] Men were bigger and stronger than women, and young men had not yet invested anything in the next generation, so they had less to lose – just like modern chimps. So began armies.

A concentration of food allowed a much denser population of people in the area, and they were better able to defend it. This increased density also allowed farmers to oust hunter-gatherers from their territory. Surely this would have happened frequently when farmers cultivated apparently unoccupied land. In short, having a higher density of people than one's competitors had great survival value, and this led to larger settlements. Populations grew from band to tribe, from tribe to chiefdom,[141] village to town, and city to state.

**********

The rising population of guarded animals had environmental consequences. In the Near East there is a crescent of mountains that curve from north Israel to south-eastern Turkey then eastwards into Iraq. The mountains embrace a lowland that was once rich in soils, rainfall, forests, game and edible plants, and is referred to as the Fertile Crescent.[142] Folk memory and luscious descriptions in religious texts record its former splendour as the Garden of Eden or Paradise. It is here that Eurasian agriculture started, and we have turned it into a fought-over desert.

Overgrazing and the excessive use of fire to create grassland from forest destroyed the root systems that bind soil particles together. Without them, summer winds blow the soil away, and winter rains wash it into rivers and thence to the sea. Plants transpire water and keep land surfaces cool. This allows rain-bearing clouds to move inland from coasts. But bare rock heats up and disperses clouds, preventing them from condensing into precipitation.[143] Disruption of a stable climate system leads inevitably to flash floods followed by drought. These simple concepts of cause and effect form an equation that much of the world's human population finds incomprehensible.[51] Both the equation and the incomprehension lie near the heart of the London questions.

**********

Seven millennia ago we began to build villages in defensible positions, and a single new language started to flood across Europe, leaving only one survivor – the Basque language.[144] This new language was brought by people who spread out from a centre north of the Caspian and Black Seas. They had domesticated the horse, and that gave them a massive military advantage. For the next 5,000 years, the horse was an effective weapon of war. It ceased to be in my lifetime with the last charge of Polish cavalry against the German army at the Battle of Krojanty on 1 September 1939.

---

[51] Personal observation from conversations with Maasai (April 2012) and many European city dwellers.

The mounted pastoralist warriors from Asia, who had abandoned their overgrazed homeland pastures, brought two-piece moulded battle-axes of arsenic-hardened copper. From the evidence of extravagant burials, they quickly subjugated the European farmers, and soon controlled large areas and many people. It is no surprise that the traditional four elements of the Apocalypse, war, famine, pestilence and death, are mounted on horses.

There is a grave beside the ruins of a Neolithic village near Heilbronn in south-west Germany.[145] It is 3 × 1.5 × 1.5 m, and it contains no grave goods, as is usual in this culture. In it are the skeletons of 34 people; half of them men and half women, people of all ages from 1 year to over 60 – an entire village. Two had been shot with arrows, the others had all been clubbed to death. None had broken forearms or other defensive fractures. None had been butchered for food as is frequently found in other Neolithic massacres. Why had these people been buried? Because their killers did not want decaying corpses lying about, for they had simply taken over the village. Were the murderers horsemen from the steppes, or merely starving villagers from nearby?[146]

We arrived in Australia 40,000 years ago.[147] The Ice Ages of the northern hemisphere had so lowered sea levels that we may have simply walked across the Torres Strait from neighbouring Indonesia; certainly it was no more than a short boat or raft crossing.[148] When we arrived, we found large marsupials grazing on open savannah with patches of forest, and they were preyed on by marsupial carnivores. Within a few thousand years we had exterminated them all except the kangaroos. When the large herbivores became extinct, the vegetation they did not eat grew into woodland which we set fire to. The conflagrations became so fierce that they gave off clouds of soot. There is a dramatic increase in soot deposits in some Australian lakes, and they can be dated to 38,000 years ago.[149] Not only did very hot fires give off soot, they also volatilised plant nutrients, such as nitrogen and sulphur. Even more damagingly, they burned the ground, destroying humus, nitrogen-fixing bacteria and the plant roots that held the soil together. Summer winds blew finer particles into river courses, and winter rains washed away many remaining nutrients. Swamps and mangroves built up around the coasts. Loss of forest cover, plant roots and soil meant that water ran off before it could evaporate back into the skies and fall as rain further inland. We converted the rich and diverse savannah of central Australia into a dead red heart.

We arrived in North America 15,000 years ago[150] by crossing the Bering Strait, and we treated the American animals as we had those in Australia. In 5,000 years 57 species were extinct.[151] In subsequent migrations to these lands by people of European descent, and whatever their declared motives, the end result has been that we have removed from Australia and North America untold quantities of food and other materials, and imported much of them into Europe. It is also true to say that there are far more people of European descent living in these countries than could live in Europe, that is, we colonised these lands to breed.

This is a biological perspective of the early and important stages of history in these continents. The events of history are repetitive: war and peace, fights and alliances, food, space and breeding. Dates, places and people change. Weapons, geography and accidents are different, but the motives seem to be always the same: the increase of territory and possessions, power and descendants.

Even though our history seems to record little but such violence, what is really extraordinary about us is how nice we are to each other. No other species of actively reproducing and territorial mammal would herd together in peaceful, appeasing, polite, smiling, patient, enormous crowds as are found every day in modern cities. We have made rules to avoid conflict. We had to because we had weapons. How did we organise ourselves to live peacefully most of the time? By cooperating. We had learned from preying on each other that larger groups are more successful than smaller ones. We needed cities. We still do, so we need to be clear about how fragile cities are. They are fragile for two reasons: behaviourally and ecologically. First, let us consider behaviour.

# Chapter 7

## VILLAGES & CITIES, RELIGION & TABOOS

The ancestral village was socially very different from a city. Although our behaviour adapted to city life, the fundamental motives driving it remained largely unchanged. Within our early communities, whether they were a hunter-gatherer family, herding band or settled village, everyone knew everyone else and had done so from birth. There were few secrets – everyone's biography was common knowledge and individuals were recognised by their faces, which is why we have such distinctive features, as indeed do chimps.[152] Secrecy was regarded with suspicion. Hence the suspicion many people feel when a woman, a soldier or a criminal hides their face.

Being known all one's life in a small community is a powerful restraint on bad behaviour. Committing a crime was a serious matter because a criminal carried a life sentence in his reputation. But the reputation of a stranger in such a community was unknown. Did he come from nearby or from further afield?[52] What tribe did he belong to? What language did he speak? What were his motives for coming to our village? Was it to steal, or reconnoitre before a raid? How did we know?

Because village raiding was a real menace, it was important for a villager to recognise first that a person was a stranger and second where he came from. We used, and still use today, obvious clues to classify strangers: face structure, skin colour and hair texture, for example, and also behavioural patterns, such as dialect,[153] accent and dress. Identifying strangers quickly had such great survival value that it led to cultural – particularly linguistic – diversity and local conformity, class division and

[52] I use 'he' deliberately because in human, chimp and gorilla societies it is generally females that join other groups, not males, as is the case in other primates; therefore, the arrival of a strange male elicits deep-rooted suspicions.

sectarianism, xenophobia and racism. So, when the first farmers spread out over Europe 6,000 years ago, they quickly evolved distinctive languages and customs. Conformity in a village was ensured by gossip, and policing carried out by all able-bodied men who knew the disputants. That is, friends restrained a man from fighting to save him from injury.[154]

The survival value to a population of it being large enough to defend or attack successfully was so strong during this formative period of our recent evolution that villages soon grew into towns and then cities. With thousands of individuals living together, people now encountered strangers whose faces they did not recognise, and they would not necessarily meet again. This meant that there was a better chance of malefactors getting away with lying, cheating or stealing in a town than in a village. Crime led to retribution and vendettas between gangs, families or communities, and this tended to break down the cohesion of a society. Clearly, such cohesion was vital when the society as a whole had to defend itself, and it was maintained by a mutually agreed, even though informal, civil code of conduct. This system worked especially well when the community was united by a common aim. For example, Alfred Wallace gave a brilliant description of self-regulation among normally warring tribes in mid-19th Century Indonesia when united by a common purpose. Their purpose was to trade and make money, and they readily understood that business flourished better when participants obeyed agreed rules.[155] Law was maintained, without any formal body of legislature, judiciary or police, by groups of citizens who met to hear what an accused person had to say in defence. They then delivered judgment and passed sentence that, if guilty, was summary and usually capital.

Once a code of conduct that ensured peace was generally accepted in the earliest cities, it required professionals who were not party to disputes to administer it. Taxes were raised to pay for them, and that meant tax collectors whose opportunities for crime tempted more than a few from earliest times, and with consequent public fury.[156]

The increasing density of people and diversity of occupations led to societies becoming more stratified than they had been among hunter-gathers. Administrators, judges and tax collectors were layers added to those of soldier and chief, merchant and farmer. Social complexity induced a positive feedback cycle: the more complex a society had become the more complex it became. People specialised in supplying the particular needs of others. They competed with each other to do so, and thereby became more efficient, which is to say that more enterprises sprang up than could survive economically, and the less efficient ones were eliminated by failure – Darwinism again. As competition increased, individuals sought fresh outlets for their services so created markets, thereby adding yet another dimension of complexity to society.

Such complexity ran more efficiently if records other than fallible memory could be kept. The Incas of South America used a system of knotted strings for this purpose, but writing was more efficient. Writing arose independently five times,[157] once in each of Sumeria, Egypt, China, Central America and Easter Island. Of these, Sumeria in the Fertile Crescent of Asia Minor was the first, and gave us the alphabet most of the world uses today. Originally writing was not meant to be understood by the mass of people; it was a clerk's code for recording debt and was also used to facilitate enslavement of other human beings.[158] Later, writing became the medium for transmitting orders of government, and then for disseminating ideas. It is only in recent centuries that universal literacy has become an acknowledged public aim.

**********

Returning to the theme of war, societies in which young men were prepared to fight were more likely to survive and increase their populations than those that had no such soldiers. Even though a soldier himself ran the risk of dying young, often before he had had children, he enjoyed some personal advantages. He had his food, clothing and pay, and (today) a sufficient pension if he survived, or one for his dependants if he did not. Such payments have largely replaced the traditional prospects of prize money or loot or even title to conquered land. In addition, a soldier gained enhanced social standing and he advertised it by a flashy uniform. Uniforms established his reputation as a vigorous fighting male and so increased his mating opportunities. From the society's point of view, young men were also cheap evolutionary material because the population impact of their deaths could be compensated for by polygyny.

Skirmishes of individuals fighting as best they could, as chimps do, gave way to more efficiently organised systems of command and thought-out strategies; and that is how we can define war.[159] Young men, trained to form a standing army, presented a problem, particularly during peace time. How did commanders control a body of ill-educated and testosterone-charged young men bearing arms and trained to kill? What was to stop them turning on the civilians they were supposed to defend, and taking over the town's women, stored goods and food? Military *coups d'états* have been common practice up to the present.

**********

There are many theories about the origins of religion, but one contributory factor seems to have been neglected.[53] As discussed in previous chapters, long after we walked erect the outer layer of our forebrains expanded suddenly to cope with our widening sociality.[160] The spare mental capacity enabled us to wonder about the world around us. We simply did not have the information to give reasonable explanations for seasons, thunderstorms, earthquakes and the like; so invoking mystical powers was the least unsatisfactory solution. We created myths, as described in Chapter 2, and became superstitious.[161] Superstition is an irrational fear of unexplained events leading to a belief in supernatural control over them. Such a belief can be fostered in others and manipulated by intelligent people for their own ends.

There was a magnificent flowering of proto-religion in the Upper Palaeolithic of south-western Europe, and it found expression in a few fabulously painted caves.[162] Additional

[53] Good pointers to this idea are to be found in Maalouf (1997), a reconstructed life of Mani (c.216–c.274 CE).

evidence indicates that shamans or first priests were sexually predatory, and this signals at least some genetic basis for the evolution of religious behaviour.[163] Scandals involving priests in the 21st Century add to this evidence.

Strong military chiefs led by example and fear, but were often succeeded by weaker sons. A clever, though weak, son preyed on the almost universal fear of death.[164] He pretended that his father was not dead, but keeping an eye on you, the army. 'My Father tells me that He will suddenly appear to chastise you, if you do not do as I say,' the weak son would say. 'I am the only person who can communicate with Him.[165] I know that, if you behave as He wants you to, you too will never die, but join Him in His new and invisible kingdom.' The son could not spread this message around his whole city or army personally, so he employed groups of ministers to act on his behalf: ministers of religion and ministers of state. Such a myth appealed to followers and in return they gave their allegiance. Superstition became so well organised and power-based that it offered prospects of self-advancement among ministers. A career structure had evolved and that turned superstition into religion.

Though this is what Jesus Christ preached against,[166] it is what Christianity became. Islam, being founded on entirely different principles, is not evolving as its popularity spreads. These are the two religions with most adherents, and I shall illustrate my arguments with reference to them, rather than attempt a wider survey.[54] Islam is highly relevant to the thesis of this book, so I devote Chapter 8 to it.

The major changes in individuals' relationships with the society in which they live are the biological processes of birth, puberty, reproduction and death.[167] Doubtless early people viewed these events as mysterious and conducted religious ceremonies to mark them. Baptism at birth welcomes the new baby into the society and gives it a name that is announced to all. If it is for medical or aesthetic reasons, circumcision takes

---

[54] Frazer (1924); this is a famously broad survey.

place soon after birth, but as a religious ritual it may be performed during puberty. In many societies, puberty is combined with a trial of competence; school examinations meet the case in Western societies. Another assessment among males is widespread when the individual has reached full height. This is when the long bones near final ossification at about 18 years, and a young man is formally admitted into manhood. In societies of European descent, there are more school examinations. In the Maasai of East Africa, young men traditionally had to kill a lion. Generally speaking, the Maasai needed to spend less time than Europeans worrying about what to do with candidates who failed this exam – the lion looked after them.

Marriage is another major step in the biology of a lifespan, and control of it by custom or law adds to the chief's or high priest's power. During the marriage ceremony, a man presents his daughter, not just to her husband to be, but also to the officiating priest or civil administrator. *Droit de seigneur* – the right of a lord to take his serfs' daughters' virginity evolved into the presentation of debutantes at the English court. Death also requires official recognition in most cultures. The chief needs to know how many subjects he has for conscription if necessary, and in many countries it is against the law to hide a death. Only recently have the rituals around marriage and death become secular.

Having its roots deep in warfare, religious rituals are still closely concerned with it. Priests are usually present to bless embarking armies, and they have a major role to play in burying and commemorating those killed in war. Whipping up enthusiasm for a just war is high on the list of religious purposes with songs such as *Onward Christian soldiers* and *Fight the good fight with all thy might*. Though priestly kings go back to the beginning of history,[168] it was St Dunstan who first anointed an Englishman, Edgar, as priest while crowning him king[169] – a ritual that persisted almost unchanged until 1953 CE. It remains to be seen whether the next English monarch will be so consecrated.

We need to be careful before attributing Dawkinsian meme status to aspects of a religion: that is to say, religious ideas becoming evolving entities in their own right. For example, at first sight it appears that a Muslim suicide bomber has been infected with ideas that have taken over his mind and superseded his gene-driven instinct for self-preservation.[170] But religious or political supporters pay money to families of suicide bombers, and that is a strong incentive to bombers' families to persuade their children to 'volunteer'.[55] Such money could well increase the chance of the family being able to bear and support more children than the one sacrificed. This signals that the driving force behind suicidal behaviour in this case is genetic as much as behavioural.

Further war-like features of religion include instructions such as: 'Fight those who believe not In God nor the last day, Nor hold that forbidden Which hath been forbidden By God and His Apostle, Nor acknowledge the Religion Of Truth, (even if they are) Of the People of the Book [Jews and Christians], Until they pay the Jizya [a tax] With willing submission, And feel themselves subdued.'[171]

This is written in the Quran, which Muslims believe contains the words of God, and which they are bound to accept as the literal truth. It is worth digressing here to look at this passage in more detail.

Muslims hold that the Quran can only be understood in the original classical Arabic in which it is written. Figure 7.1 is a copy of the original that is translated in the previous paragraphs.

---

[55] Searching for 'payments to suicide bombers' on Google yielded about 4.5 million references in October 2015.

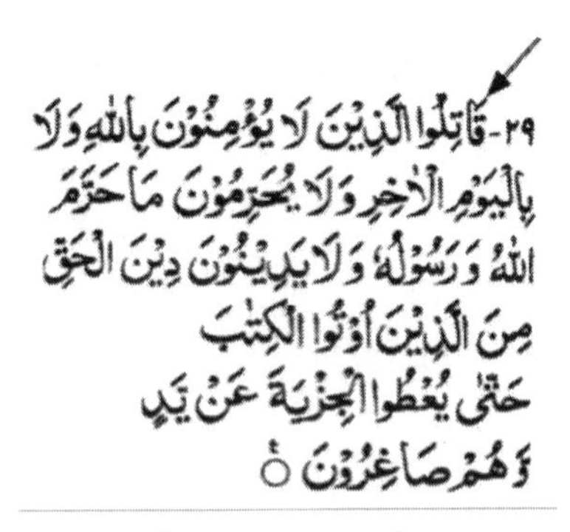


**Figure 7.1** The 29th verse of the 9th Sura of the Quran.

An alternative translation by Khurram Murad is quoted by Sikandar:[172] 'Kill those who believe not in Allah nor his Last Day, nor hold that forbidden which hath been forbidden by Allah and His Apostle, nor acknowledge the Religion of the Truth, even if they are of the People of the Book.'

I have discussed this with an Arabist,[56] who wrote: 'It [the first translation given above] is an accurate translation but the first word 'QAATILU' [indicated by the arrow in Figure 7.1; the right-hand two symbols of that line are '29', and, being figures, they are read from left to right] which is the imperative of the third form of the verb Q-T-L means 'fight' and not kill. The first form, which lacks the emphasis on the first vowel, means 'to kill'. Arab verbs each have a potential 10 forms where the 3 radicals are added to or rearranged. Each form is recognised by its specific pattern and has a connected though different meaning. So what you have in the Arabic Qur'anic Verse is 'Fight those who believe not in Allah' NOT 'Kill' [original emphasis].' Dr Badawi, an Islamic leader in London whom I consulted in 2004, concurred.

When travelling in Morocco (1990–2007), on several occasions I asked hoteliers, taxi drivers, guides and others to read the first line of the above Arabic 9/29 and to tell me what it meant. Without exception they (and I cannot remember how many they were) told me that *qaatilu* meant 'kill' or '*tuez*', if we

[56] Michael Baddeley

were speaking French. Several emphasised the point by drawing their right index fingers across their throats.

My Arabist friend went on to write: 'Islamic extremists and fundamentalists may be tempted to replace 'fight' by 'kill' since they frequently find atonement for their violence by quoting verses from the Quran as a justification and get it purposefully wrong.'

**********

It is interesting to compare the different ways in which people are recruited into each of the three great monotheistic religions. If their parents professed a religion, it was assumed that a child would have been indoctrinated from an early age;[173] to become a Jew takes months of instruction and examination, Christians are admitted with perfunctory questions, but Muslims vow to fight, or even to kill, those who do not say that they subscribe.

Having a god on one's side in war is reassuring. The trouble is that both sides believe it. In World War I officers in the European trenches spoke freely of *God* and *Gott* as opposed tribal deities.[174] In a room off the courtyard beside the cathedral in Cusco, Peru, there was, in 2006, a model of the great Inca temple that had stood on the site before *los conquistadores* obliterated it with their Baroque church. In one corner of the model a small group of citizens and priests were gesticulating in front of a dozen or so gold figures that were obviously their gods. Behind and to their right was a small chamber with tiny figures standing around. I asked the guide what they represented and he replied, 'That? That is the prison for captured enemy gods.'

Once religion was established as a way of controlling armed young men, the priesthood that managed them needed an income. Establishing a theocratic society, as has been done in some Muslim states, effectively eliminates opposition and establishes the priesthood in permanent power with its associated wealth. In more liberal societies, such as the Anglican Church, the post of parish priest in England was called 'a living', and it was just that. An effective way of maintaining such a living is to command '*Love thy neighbour* ….' Though love appears to be an exhortation to increase altruism,[175] its practical application is less sure.[176] Those who live off the fruits of altruism need to encourage it in others, or their source of fruit will fail.

Forgiveness is applied more to children than to adults; bestowing forgiveness on adults tends to reduce them to the status of children, and this makes them more biddable. The golden rule to *do unto others as you would have them do unto you* has a far-reaching effect on society.[177] All but a few people recoil at the thought of their being executed or even imprisoned, therefore they hesitate to recommend such treatment be inflicted on others, however justified such a sentence may appear to others who do not have to take the decision. Thus law breakers may not be punished severely enough to prevent them from offending again. Also, if a priest's living depends – at least to some extent – on forgiving people, he needs a supply of penitents. In a parallel way, it would cause some financial difficulty to local politicians if every motorist obeyed the speed limits.

Chastity, poverty and obedience also have their place in religious rules. Chastity weakens family bonds in that the chaste are supposed not to have children. Being a father increases the DNA-driven tendency to self-preservation in soldiers,[178] which is not good when planning military strategy. Chastity in men, as well as the chance of their being killed in battle, makes more women available for rulers – a situation that Uriah the Hittite discovered.[179] Dictating periods of temporary chastity within marriage, in the knowledge that the men so restricted would be tempted to seek sex elsewhere, reduces the likelihood of legitimate heirs, and therefore increases the chance of testamentary gifts to religious bodies while not affecting the overall birth rate.[180] Insisting on poverty tries to keep the minds of the poor on spiritual, not worldly pursuits, so keeps them poor. It is also a way of increasing revenue by donations from those who feel guilty about not complying.[181]

Obedience is for control; rulers do not like the idea of people making decisions on their own, even though those people may want to think for themselves. The application of such diverse thought may well lead to innovation and increased general wealth;[182] however, this is of small concern to those who seek only to widen margin between their own wealth and that of others. The concept of sin emerges from these strictures by giving people rules that go against their natural inclinations. The internal conflict makes them unsure of themselves and they then find it difficult to take decisions. Sin is transgressing a law or social norm.[183] Sinning makes sinners feel guilty, lowers their self-esteem and makes them more willing to do what others tell them to do. At some time in our lives, most of us want to do things that are unacceptable to others, and our conscience (or other-consciousness) restrains us – unless we think that we can remain undetected. Continually looking over the shoulder makes us insecure. A person is vulnerable to manipulation when they suffer internal conflict between their personal feelings and social rules imposed on them.[57] Of course, their escape is by understanding the evolutionary origins of feelings and rules.

There is little point in having a god that is not omnipotent, and Islam's fatalism depends on it. An MIT-trained professor was explaining the tectonic causes of the 2005 Kashmir earthquake to a class of graduate physicists in Islamabad. When he had finished, hands shot up all over the room. 'Professor, you are wrong,' the students said, 'that earthquake was [caused by] the wrath of God.'[184] Instilling a sense of fatalism is an important element of control. If people can be convinced that whatever they do has no influence on the future, rulers and priests have control over them. In this connection, priests declare that they are the sole conduit between their god and the people (see Chapter 8).

To persuade people that death is an illusion that they will not really die, is powerfully attractive to most people. Rewards in

[57] Romans 7.18–19: 'For I know that in me (that is, in my flesh) dwelleth no good thing: for to will is present with me; but *how* to perform that which is good I find not. For the good that I would I do not: but the evil which I would not, that I do.' Quoted by Wilson A. (1997: p. 230), brackets and italics in the Authorised King James Version. See also Wilson's superb note on Romans on p. 195 in his book.

the next world are certain, priests say, if the poor remain poor and obedient. The word *paradise* has its roots in the name of the gardens[58] outside Babylon and Samos where prostitutes could be found.[185] This is the origin of male Muslims' obsession with sex after martyrdom, and with virgins in particular. Having sex with virgins increases the probability of paternity and reduces the risk to the man of contracting a sexually transmitted disease. Paradise, or any other place of sublime happiness, is a myth used by rulers and their priests to tempt people into overcoming their natural fear of death, and this makes them better soldiers. As I shall argue later in this book, there *is* an afterlife but it has nothing to do with gods and priests.

Many religions exhort their followers to feed and breed so that the population goes on growing. This not only increases the population for defence or offence, but also provides a living for rulers and priests:

> 'And God blessed them [Adam and Eve], and God said unto them, Be fruitful, and multiply, and replenish the earth, and subdue it: and have dominion over the fish of the sea, and over the fowl of the air, and over every living thing that moveth upon the earth.
>
> And God said, Behold, I have given you every herb bearing seed, which is upon the face of all the earth, and every tree, in which is the fruit of a tree yielding seed; to you it shall be for meat.'[59]

Looking at the land that was the cradle of both Eurasian agriculture and the three principal monotheistic religions, it would appear that mankind has obeyed this instruction and, in so doing, turned the Garden of Eden into a desert. There are still

---

[58] The Quran is most insistent on heaven as a garden: see 3.15, 9.72, 14.45, 18.31, 22.23 among some 30 references.

[59] Genesis 1.28–29 (my brackets) which serves for Jews and Christians. There are many similar exhortations to the same ends in the Quran in the translation by Dawood (1956); see his index.

some fish in the sea and rainforests to subdue and have dominion over; after that, then what?

In sum, it seems likely that farming, war and religion evolved hand in hand about 10,000 to 8,000 years ago. We have identified superstition and ignorance, greed and fear as deep-seated motives in the origin of religions, and they are not good foundations for answering the London question, nor for planning survival through the ecological apocalypse that faces us and our children.

**********

Another impediment to understanding that is closely related to religion, is the concept of *taboo*. A definition of 'taboo' is 'adjective ... set apart for or consecrated to a special use or purpose; restricted to the use of a god, a king, priests, or chiefs, while forbidden to general use.'[186] Though this definition of taboo is precise, I shall use the word more loosely. I shall include in its meaning all inhibitions that range from 'not quite nice' through to 'non-PC (politically incorrect)', even 'outrageous', 'disgusting', 'inhuman', 'sacrilegious', 'blasphemous' and 'illegal'. Suitably manipulated, taboos are useful devices for maintaining a position of social or political power; and we see such abuses in flagrant operation today. How often do you hear or read, 'It is not in the public interest to reveal ...'? Here are some examples of taboos.

Parasites have imposed powerful natural selective pressures on the evolution of species. Indeed the theory that sexuality itself arose as a defence against parasitical attack was the second of the evolutionary biologist W. D. Hamilton's theories that changed the way we think about ourselves.[187] In such parasite-driven regimes, we find that more complex animals have evolved patterns of behaviour that reduce the risk of infection. Such behaviours include: being particular in selecting foods; close attention to hygiene; strict regulations regarding outbreeding; and measures to eliminate obvious vectors of parasites. For example, eating the flesh of *Sus scrofa*, the domestic pig, is taboo in Judaism and Islam. Such a prohibition makes sense in primitive societies because we humans are host to a parasite called *Taenia solium*, the pork tapeworm. The adult reproductive stage of this animal lives in our intestines, while its eggs leave our bodies in our faeces. Pigs feed in soil where our dung is likely to have fallen, and are a traditional prey animal of mankind. A larval form of the tapeworm develops in the pig's muscle, which is the part of the animal we prefer to eat. Drains and sewage works in developed societies have made this taboo irrational.

Taboo on handling food or holy books or making contact with others without washing is a sensible precaution against transmission of parasites. Forbidding the use of the left hand for eating, but reserving it for personal sanitation helps to break the direct life cycle of ascarid worms. Taboo on marrying outside the group of people that share these behaviours has three principal functions: it reduces the chances of importing sexually transmitted diseases, reinforces group identity and adhesion, and keeps wealth within the group.

Interestingly, some groups of people, who are united by intense taboos, hate and wage war with savage ferocity on others that share most of these behaviours but deviate in some small particulars – at least the particulars seems small to outsiders.[188] Adherents of Sunni or Shia versions of Islam spring quickly to mind today, and Catholics and Protestants still have their differences, especially in Northern Ireland. Ritual distinctions between Jews and Muslims are, in fact, fundamentally slight, yet their mutual hatred is legendary. Indeed this hatred is so profound that the declared intention of many adherents is to eliminate rival deviant groups – a program that often amounts to genocide. Such savagery is understandable in biological terms: the more similar two species, races or beliefs are, the more alike are the resources on which they depend (food and space in species and races, or potential followers in religious sects), therefore, the more fiercely the divisions compete and try to eliminate one another.

The fact that these deeply entrenched individual behaviours also appear in groups of animals other than mankind suggests that they are genetically coded for. If such instinctive behaviours were genetically linked together, I would be tempted to label them with the old-fashioned term ‘supergene’. When asked whether he thought that there were possibly a gene for religion, W D Hamilton replied that it was very likely that such a gene or group of genes existed, and that he would expect to find it most highly evolved in ants.[189] I have more to say about Hamilton in later paragraphs.

**********

Here are three more sets of ideas that were once taboo. All three were exploded by naturalists.

The earliest humans to think about it, believed that the Earth was fixed and that the heavens rotated around it.[190] They saw that the Sun, the Moon and five smaller heavenly bodies moved relative to the fixed pattern of stars. The Sun created day, and the Moon dominated the night, and also seemed to control the tides and women’s menstrual cycles, so the Sun and the Moon soon became invested with supernatural powers – gods. By association, the other five wanderers (Greek: *planetes*) were also attributed divine powers.

The ancient Greeks tried to rationalise the movements of the Sun, Moon and planets by pure logic, but their explanation did not accord with observations. Aristarchus (c.310 to c.230 BCE) speculated that Earth rotated about the Sun, but his ideas did not catch on. Later, the Earth as the centre of all things meshed well with Christian teachings, especially in the 13th Century writings of St Thomas Aquinas. By the 16th Century, geocentrism had become religious dogma and alternative ideas taboo.

Nicolaus Copernicus (1473 to 1543), a canon of Frauenburg Cathedral in Poland, studied astronomy and, in 1514, privately circulated the suggestion that the Earth rotated on its own axis and also orbited the Sun. Certainly the observations and calculations of others made more sense if explained this way. His friends urged him to publish, but he held back for 25 years. There were further delays because of opposition from Martin Luther and other reformers, but Copernicus lived just long enough to see the first printed edition of *On the Revolutions of the Celestial Spheres* (1543). In 1613 Galileo Galilei published evidence in support of Copernicus. Three years later

Copernicus's books were banned. Galileo was tried for blasphemy in 1633 and sentenced to house arrest for the remaining eight years of his life.[191]

Questioning that the Earth was the centre of the universe shook established religious beliefs. Eventually most Christians conceded that Copernicus and Galileo were right, but insisted that *Homo sapiens* was specially created in the image of their god, and discussion of *that* was taboo. The story of the demolition of this taboo by Charles Darwin is topical today with the rise of special creation remarketed as 'intelligent design'. I sketch the background in Chapter 4. Eventually, more rational theologians conceded that Darwin was right: we are descended from ape-like ancestors that also gave rise to apes; but theologians insisted that human moral sense was given by their god, and that denial of *this* was taboo. Wallace also believed it. Indeed, Darwin himself could not understand morality and self-sacrifice. His theory of natural selection depended on each individual acting in its own interest. Why then do many species – most notably our own – appear to behave altruistically? Darwin expressed his worries thus, 'I will … confine myself to one special difficulty, which at first appears to me to be insuperable, and actually fatal to my whole theory. I allude to the neuters or sterile females in insect communities: …'.[192] In fact his worry was not so much about the apparently altruistic behaviour of sterile worker ants, wasps and bees (collectively called hymenoptera), but the fact that they were there at all.

Less well known is the story of another brilliant naturalist, W D Hamilton (1936 to 2000) whom I mentioned earlier, so I shall give him some space now. Anyone who has been stung by more than a thousand hymenopterous species is worth listening to.[193] After school,[60] Hamilton read the Natural Sciences Tripos at St John's College Cambridge with genetics in part 2. He failed to get a First Class degree, almost certainly because he understood

[60] Hamilton and I overlapped at the same school for two years, though we never knew each other. I have compiled a short account of that time, and you can read it at http://evolution.unibas.ch/hamilton/img/hamiltonschool.pdf. See also Segerstråle (2013: ch. 3).

the mathematics of population genetics better than his examiners did.

Hamilton had hit on an idea that no one else had thought of – or articulated clearly – and he wanted to pursue it. He was rejected for research at Cambridge because thinking about evolution there in those days was still of the 'good-of-the-species' kind. He failed to get a job teaching in schools because he did not have a degree in biology, the interviewers said. Depressed, lonely and contemplating a career in carpentry,[194] he persisted in his search for the chance to work out the problems that interested him. Eventually, he found a supervisor at the London School of Economics and a desk at University College London.

Hamilton asked himself why birds give alarm calls.[195] It cannot be an advantage to the individual that makes the call because it draws the predator's attention to the caller while others flee in silence. Hamilton reasoned that most birds live in flocks of individuals that are related to each other with varying degrees of closeness. In other words, they share genes.

Before going further we need to be clear what 'sharing genes' means. All living things share some genes, and individuals within a species share all but a few. It is the proportion of these few genes to the others that reflects the closeness of family relationship. Accordingly, you share half these few genes with each of your brothers, sisters, parents and your children, a quarter with each of your grandparents, uncles, aunts, nieces and nephews, and an eighth with each of your great-grandparents and first cousins.

To round off this argument, we need to introduce another set of ideas which I will deal with more thoroughly in the next chapter. It is the thought that genes are replicators; which is to say that the reactions they undergo give the impression that it is their purpose to make copies of themselves. 'Purpose' is a word loaded with human connotation, if not emotion, that is inappropriate here. Nevertheless I shall use it, even though genes are no more than mere chemicals. Likewise we can estimate a gene's 'success' (another loaded word) by the number of copies of it that results from what it does. If this is a good description of reality, then the information carried by the gene is more important than the actual assemblage of atoms that store that information, which is like thinking that the ideas in a book are more important than the paper it is printed on. If these presumptions are reasonable, then genes can ensure the survival of the information they carry in two distinct ways: they can make copies of themselves, as described in Chapter 3, or they can cause the individual living thing in which they find themselves to behave in ways that protects copies of the information that is being carried in another individual. Therefore, if a particular gene could induce the individual animal that carried it to give an alarm call that cost the caller its life but saved three siblings or offspring, or five nieces or nephews, or nine first cousins, it would have fulfilled its 'purpose' of replication. That is why birds give alarm calls. The genes are manipulating the individual bodies they are in to the benefit of the genes themselves – not necessarily to the body that carries them.

It is not easy to work out gene frequency in birds so, perhaps inspired by Darwin's solution to his own difficulty,[196] Hamilton asked himself why worker bees or ants work for the queen – it certainly does not give them a reproductive advantage because they are all sterile. He realised instead that they are helping to propagate the genes they share with their mother the queen ant. In other words, natural selection can operate on related groups of individuals as a whole. In effect, ants reproduce only by young queens founding new colonies, and bees by swarming. In these insects, laying eggs is functionally equivalent to the cells of vertebrates dividing.

There is a further complication with birds' alarm calls and it has to do with their social organisation. The complication is more clearly demonstrated by the behaviour of suricates, which Hamilton quotes.[197] Suricates, sometimes known as meerkats, are engaging little mammals found in arid parts of southern Africa. They live in family groups, digging a communal den, and foraging together. Usually, it is only the dominant male and dominant female within a family group that breed. Suricates have very good eyesight and rely on it to detect predators. Most families post sentries – an individual that climbs to a higher place where it can see far, and calls in alarm when danger threatens. Clearly such exposure and loss of feeding or resting time is a significant disadvantage to the individual sentry, yet family groups that post sentries leave more offspring than families that do not. Maybe those individuals that have served longer as sentries rise up through the social hierarchy faster and so have more chance of breeding later in their lives. If so, this is a case of deferred reward for a seemingly altruistic service. Nevertheless, it is generally true that more offspring survive in groups of social animals in which individuals cooperate than in those that do not. No animal species has benefitted so much from cooperation as our own, and it has led us to our present population difficulty.

A few years after submitting it, Hamilton's PhD thesis was reluctantly accepted, and eventually published in three papers: *The Evolution of Altruistic Behaviour*, and *The Genetical Evolution of Social Behaviour, Parts I & II*. They formally show that there is an evolutionary advantage in cooperation, thus filling the biggest gap in Darwin's contribution. These two papers are among the most quoted in all biology, if not all science.[198] They are the foundation of E. O. Wilson's *Sociobiology* and of Richard Dawkins's *Selfish Gene*, and opened up vast fields of understanding in widely differing disciplines. Hamilton's explanation shatters, not only the view that our moral sense was given by a god, but also our concept of self. It is a key to the real nature of individuality, and opens the door to a way we can go forward with confidence from the obvious threats of today.

Taboos are still being cultivated by rulers with the purpose of restricting human enquiry – enquiry that might question the rulers' position and associated wealth and mating opportunities. In many Western societies, the most powerful taboo operating today forbids us to question the sanctity of human life. Yet believing life to be sacred seems to be at the centre of our population explosion. Before we can face that problem there are several other concepts that need clarification. Chief among them is the idea of 'I' – the individual; and that is intimately connected with religion. Before dealing with that, I feel (*sic*) obliged to deal with a particular threat to world stability. Its avowed intention is to breed maximally, yet it has scant regard for individual life. This makes it quite different from the main stream of human concern. It is Islam.

# Chapter 8

## THE SPECIAL CASE OF ISLAM

If we accept, as a broad generalisation, that most of us want to breed but do not regard it a duty to have as many children as we can, and that we hold all human life to be sacred, then Muslims are different. They do regard maximal breeding as a duty, and they do not hold *kaffir* (non-Muslim) life as sacred as that of believers.[199] These are contentious claims, and, while inviting contradiction, I shall support them with evidence and argument. The first pieces of evidence I offer lie in the answers to five questions and their subsidiaries:

- Do you owe your allegiance to Islam and Sharia[61] law above what you owe to the law of the land you live in?
- Do you believe that the ideas in the Quran, as expressed in classical Arabic, are absolute and not open to common commentary?
- Do you believe that kaffirs have different political and social rights from Muslims?
- Do you believe that all people should eventually become Muslim and obey Sharia law?'
- Do you believe that the life of the Prophet Muhammad is the model for Muslim men to follow?[200]

If your answers to all these questions are positive, you are a Muslim; but if any one of them is negative, I ask you to write down the particulars in which the Quran, Sharia law or the life of Muhammad may be interpreted. If you allow that interpretation, how much further may anyone go in modifying Islam? Would you go so far as to allow the interpretation that I offer below? If not, who is to decide that I have gone too far?

[61] I have rendered Arabic words in the English alphabet without accents.

I shall deal only with the last of this chapter's opening questions, and do so by describing then analysing the life of Muhammad with a naturalist's eye. On the basis that it is as important to know how one knows something as to know it, I give my main source as Ibn Ishaq. He was born in Medina, son of a freed slave, in about 672 CE. He studied apostolic tradition, attended lectures in Egypt and was highly thought of on his return to Medina. This lasted until he attracted criticism from powerful men, whereupon he moved to Baghdad where he died in his hundredth year. He based his account on a number of previously published books, and the testimony of witnesses who had second- or sometimes first-hand knowledge of Muhammad. I read Ishaq in Alfred Guillaume's 1955 translation. In his *Introduction*, Guillaume gives details of Ishaq's scrupulousness about giving the sources of his information. Thanks to his record and the oral tradition that quickly grew, more is known about the founder of Islam than of any other major religion. I give other sources in the endnotes, which often indicate paragraph as well as page number.

## A brief life of Muhammad

### *Background*

During the 6th Century CE, two empires dominated the Near East: the Byzantines with their capital in Constantinople and the Sasanians (Iranian) centred on Baghdad. In Egypt, the Ptolemaic dynasty ruled a lesser and largely Christian population. None of these powers held dominion over the Arabian Peninsula, which was inhabited by several Semitic tribes. These tribes included monotheistic Jews and Christians (referred to in the Quran as 'the people of the book') and polytheistic pagans. Some tribes grew crops and others were nomadic pastoralists. They traded with travellers, which they often robbed and killed, and were almost permanently at war with each other. A crime committed against one individual was avenged by any member of the offended tribe, clan or family. Some semblance of peace was maintained by the fear of prolonged vendetta.[201]

Mecca was a trading town on a desert crossroads between Asia, Europe and Africa. It was dominated by a large polytheistic pagan tribe called the Quraysh (sometimes written 'Koresh', just as 'Quran' can be written as 'Koran'). The Quraysh were divided into a number of clans, each containing several families.

Mecca also contained an important pagan shrine called the *Kaaba* (cube). The Kaaba housed a sacred black stone[202] that is, in fact, a meteorite 20 × 15 cm long, which was thought to have a message from the gods inscribed upon it.[203] This black stone was associated with a pagan god called *Allah*, who was the preferred god of the Quraysh. The Kaaba also contained idols of three pagan goddesses.[62]

Nearby was a well called Zamzam from which the local clan, the Hashemites, sold water to pilgrims. Biblical legend has it that Hagar, an Egyptian maidservant of Sarah, Abraham's wife, was given by Sarah to Abraham to bear a child. The product of the union was Ishmael, Abraham's firstborn. When Sarah gave birth to Isaac, her jealousy of Hagar and Ishmael reached a climax, forcing Abraham to cast Hagar and Ishmael out into the wilderness. Having exhausted their water supply, Hagar and Ishmael miraculously came across the spring of Zamzam. While Ishmael founded the Arab people,[204] Isaac founded the Jewish people.[205] The well of Zamzam was later covered over and lost, then rediscovered in the 6th Century CE by Abdul-Muttalib,[206] who was head of the Hashemites. In 570 CE Abdul-Muttalib's eldest son, Abdullah, married Amina and had a son they called Muhammad.[207]

## *Early years*

It was a year of famine and Amina had no milk, so a wet nurse in a Bedouin tribe fostered Muhammad. Abdullah died poor

---

[62] The gods of the Nabataeans were represented by rectangular blocks and a black stone. Three of their minor female goddesses were called Allat (the female form of Allah), al Uzza and Manat – the same as those in the Kaaba. (Taylor (2012: pp. 124, 128))

when Muhammad was a baby, and Amina died when he was six. In the next two years, Muhammad herded sheep and had at least one epileptic fit.[208] He returned to Mecca, when he was eight years old, to the care of his grandfather, Abdul-Muttalib, who died soon after.[209] Muhammad then lodged with his uncle Abu Talib. Later, Abu Talib took the boy to Damascus in an unprofitable trading caravan. There they met a Christian monk who influenced Muhammad.[210]

### *Mecca*

When Muhammad was 20, war between two tribes broke out during the sacred month, when fighting was prohibited,[211] and he fought in it briefly. In more peaceful times, he worked in a store selling agricultural products[212] and built up a good reputation. Khadija, a 40-year-old widowed business woman employed him to lead a trading caravan to Syria, and he did this profitably.[213] She married Muhammad when he was aged 25.[214] Ten years later, Muhammad appears in legend as one who resolved a tribal dispute over the rebuilding of the Kaaba.[215]

He realised that fighting between tribes discouraged pilgrims to the Kaaba, and this diminished water trading for his Hashemite clan. Muhammad also realised that the monotheists resented the polytheist's pagan idols in the Kaaba.[216] He resolved to found a new religion with a supreme god everyone would accept. For the next five years, he wrestled mentally with the challenge, spending much time walking alone in the hills around Mecca. He told Khadija that he had visions of the Archangel Gabriel who brought messages from a god he called Allah, and that he, Muhammad, was the messenger and the last prophet. In another such revelation, Muhammad had an erotic vision in which he could not make out whether he was in contact with Gabriel or Satan. Khadija participated physically and resolved the matter in Gabriel's favour.[217]

In 610 CE, when Muhammad was aged 40, he began his mission by converting some of his friends and relatives to his new monotheistic religion he called *Islam* – submission.[218] He preached aggressively against polytheism, claiming that what he said was the word of Allah, the one true and only god, and revealed only through him. This deeply offended the majority of the Quraysh and they persecuted him.

### *The Satanic Verses*

Realising that he was driving away possible converts to his new religion, Muhammad sought a compromise with the Quraysh, and waited for a message from Gabriel. It came and he said it agreed that, if the Quraysh would worship Allah, Allah would allow that the idols of their three pagan goddesses could intercede on behalf of worshippers. This greatly pleased the Quraysh. Gabriel then appeared again to Muhammad and berated him for propagating something Allah had not said.[219] Muhammad was horrified that he could have misunderstood, and recanted the first revelation.[220] It was suggested that Satan had forced him into this error. That is why these *suras* (chapters) of the Quran, which were deleted from it, are called the *Satanic Verses*.

Muhammad continued to preach about his increasingly unlikely visions, including a journey to Jerusalem, ascension into heaven and return to Mecca, all accomplished in one night.[221] His then wife, Aisha, later reported that he was at home all that night, so the vision was probably a dream.[222] This is the basis for the present Arab-Israeli conflict in Jerusalem, and is the root reason why Muslims control the Dome of the Rock today.

Khadija and Abu Talib both died in the year 619 CE.[223] Without Abu Talib's protection, the Quraysh increased their attacks on Muhammad and eventually drove him out of Mecca. This move, in 622 CE, is taken as the beginning of the Muslim Era and is recorded as 0 AH (*anno hijrae*, or erroneously, *anno hegira*).[224] The Muslim year is 12 lunar months long, so it is 11 days shorter than the year the rest of the world uses; this is why the month of *Ramadan* advances in the CE year. Muhammad took refuge in Medina where many of his converts had preceded him.

### *Medina*

In Medina, Muhammad made friendly contact with the Jewish tribes.[225] He rejected their clapping and trumpets to announce prayer, and instituted instead the call of Bilal Ibn Rabah, a freed slave who was one of his earliest converts. Bilal had a hugely powerful voice, and was the first *muezzin.*[226] Even so, Muhammad adopted many Jewish religious rituals.[227] He followed the Torah's prescription of stoning adulterers to death, and relished ordering it.[228] He invited Jews to convert to Islam but they repudiated him, and relations between them and him deteriorated.[229] Like the Jews he prayed facing Jerusalem but after the rift with them, he changed the direction towards the Kaaba in Mecca.[230]

Muhammad was described as a small man who caught the eye. He opened his huge mouth like a crocodile when he laughed, which was not often. He could be agreeable and rather boisterous company, and walked as if he were perpetually descending a hill. He used *kohl* on his eyelids, dyed his beard red or yellow, and favoured brightly coloured clothing.[231] He had fine speech and beautiful diction,[232] but was probably illiterate.[233]

His revelations, which he said were transmitted by Gabriel from Allah, were collected together, largely from oral tradition, and written down during the first two centuries AH as the Quran. In reality, the Quran is probably a precipitate of social and political prejudices of that time, and collected together by several different authors.[234] If the Quran was the word of Allah, why does it include several contradictory revelations, and also the phrase, 'If Allah wills …'?[235] It is also strange that Allah made Muhammad both illiterate *and* the conduit for divine revelations.

After Khadija's death, Muhammad gave full vent to his huge sexual appetite, for which he was admired and envied by his followers.[236] He had nine official wives[237] and several verses of the Quran are devoted to their bickering.[238] He married Aisha when she was six years old and consummated the marriage when she was nine.[239] He was then in his early fifties and fascinated by all sexual matters, including whether the features of a child were determined by the man's or the woman's 'semen'.[240]

## *Jihad*

Within a year of arriving in Medina, Muhammad began raiding tribes who would not convert to Islam,[241] particularly the trading caravans of the Quraysh of Mecca. He made 77 raids in the 10 years he was in Medina,[242] and Ishaq describes 45 of them; several were during the sacred months.[243] Raids followed a uniform pattern. Nearly all of the men were killed, either in battle or by beheading after they had been captured; Muhammad delighted in the execution of prisoners.[244] Women were divided among his followers, with Muhammad taking first pick,[245] and incorporating some into his harem, though not always. If any displeased him he had them killed brutally.[246] One had her legs torn apart between two camels, and her daughter given to the soldiers.[247] Children were enslaved and treated as other booty, which was shared out after Muhammad had taken 20% for himself.[248] Apart from this and the tribute he extorted from conquered people, he appears to have had no other source of income.

He inspired suicidal devotion in his soldiers by promising them the delights of Heaven (paradise) after death or the fear of Hell.[249] Both these are described in the Quran in some detail.[63] Islam's concept of a martyr is not a person that suffers for a cause, but one who dies while killing a kaffir.[250] Muhammad's followers were convinced that opposition to him or Islam was worse than breaking agreements and laws. 'Good' was anything that benefitted Muslims and 'evil' anything that hindered them.[251] Allegiance to Islam was more important than to family.[252] Today Muslims claim never to kill or injure anyone who is not guilty. In their eyes, everyone who is not a Muslim is guilty.

---

[63] In the index to his translation of the Quran, Ali gives a page of references to Hell (p. 1833) and one-and-a-half pages to Heaven (p. 1832), with special emphasis on women: xxxvii (41–49), xxxviii (49–52), xliv (55) and lv (55), though he calls them 'chaste', rather than 'virgins'.

Throughout Ibn Ishaq's biography of Muhammad, descriptions of raids, battles, murder of male prisoners – usually by beheading – are too abundant to catalogue; indeed, they constitute the bulk of it. Even so, it is thought that Ishaq's biography probably cuts out the worst atrocities.[64] Though Muhammad's original aim appears to have been to establish a monotheistic religion, it quickly became subverted, at least in the minds of his followers, to simple booty gathering – including women. With a dedicated army who were convinced of either earthly riches or a glorious afterlife,[253] and plenty of money to buy arms, Muhammad quickly conquered the Arabian Peninsula. In 9 AH, he attacked the Byzantine Empire.[254]

### *Death and succession*

Muhammad died in 12 AH (632 CE), when he was 63 years old.[255] Abu Bakr, who had supported Muhammad throughout, and whose daughter Aisha was one of Muhammad's wives, took over leadership of Islam under the title of *Caliph*.[256] There was some opposition to start with, but it was soon put down by force. Abu Bakr was succeeded by Umar ibn Abd al-Khattab. Within a few years, the Islamic armies had conquered the surrounding Byzantine and Sasanian empires, which were exhausted by fighting each other.[257]

The centre of Islamic political life began to move from the deserts of Arabia to the fertile plains around Baghdad. Umar was murdered in 644 CE and a faction in Medina, seeing the base of power slipping eastwards, appointed Uthman ibn Affan as third Caliph. He was murdered in 656 CE.[258] Ali, Muhammad's cousin and son-in-law, was from the Quraysh tribe, and set up as Caliph in Kufa. His rivals were based in Basra, and he defeated them in battle. The next challenge was from Muawiya ibn Abi Sufyan, governor of Syria. They fought but negotiated a truce. Members of both sides had difficulty accepting this truce because each was certain that an invincible Allah was on their side, and to accept a truce was to admit

[64] Indeed, Spencer (2006: p. 28.2) writes: 'Ibn Ishaq's life of Muhammad is so unashamedly hagiographical that its accuracy is questionable.'

Allah's fallibility. Ali was murdered in 661 CE and Muawiya was declared as the fourth Caliph in Jerusalem, which was then part of Syria. Muawiya built the Dome of the Rock on the site of the Jewish temple.[259] With the expanding empire, the capital moved from Medina to Damascus, whose land could better support an administration and army.[260]

These first four caliphs are accepted by all Muslims as 'the rightly guided', even though three of them were murdered, presumably by Allah's will. When Muawiya died in 680 CE, a family descended from a man called Umayya, and thus called the Umayyads,[261] ruled for a few generations before civil war between factions, with each faction insisting that their families should inherit the post of Caliph.[262] The matter was further complicated by two issues. The first was local resentment between two separate classes: conquerors and conquered, and between long-term converts and newcomers. The second issue was simply the expansion of the empire, which by 711 CE, stretched from northern India to the gates of Vienna, and then through North Africa and Spain to the Pyrenees.

### *The Sunni/Shia division*

Muhammad's words and deeds are the highest pattern of conduct and form the only absolute standard within Islam.[263] They are called *sunnah*, and include the traditional ways of electing leaders.[264] Followers of this belief are referred to today as Sunnis. In 7$^{th}$ Century CE (1$^{st}$ Century AH), some Muslims believed that the genetic descendants of Muhammad should take precedence because they were born with a special insight into the meaning of the Quran.[265] These dissenters faced the problem that Muhammad's sons died young, so they chose his cousin, Ali, as his genetic heir. Having married Muhammad's daughter Fatima, Ali was also Muhammad's son-in-law. Husayn, Ali's son by another woman, moved to Karbala, in what is now Iraq, where he was killed. This remembered martyrdom inspired a faction to form what became known as 'partisans of Ali', *Shiat Ali* or *Shias*.[266] Today, they represent about 15% of the 1.6 billion Muslims worldwide.[267]

### *Muslim achievements and al-Ghazali*

In spite of squabbles over leadership, the Muslim empire experienced a magnificent cultural flowering over the next three centuries. In it, architecture and geometric art, mindful of Muhammad's prohibition of representational figures, explored new dimensions. Mathematics was perhaps Islam's greatest contribution to global civilisation, for they introduced to the Western world the Hindu concept of zero and the symbols for the integers 1 to 9, which are used universally today. These mathematical notations are vastly more efficient than Roman numerals and enabled superb Islamic architecture; they also allowed mariners to explore the world known to Europe and push its boundaries eastwards.

Advances such as these required original thought, which is not conducive to blind obedience to a violent creed. By the 11th Century CE, pressure from religious clerics, notably Abu Hamid Muhammad ibn Muhammad al Ghazali, had made denial of causality part of Arab culture. Thus reality, other than survival, was rapidly becoming inaccessible to Muslims.[268] The library at Cordoba contained 400,000 manuscripts – possibly more than in all of Western Europe; it was destroyed by Muslim Berbers in 1013 CE.[269] Today, more books are translated into Spanish in one year than have ever been translated into Arabic.[270] Innovation is a high offence in Islam.[271] Hence, original thinking is eliminated from Islamic education, which is reduced to mere mimicry.[272] Every discussion about a theory that the Prophet did not discuss is an error.[273] All nature is miraculous and all miracles are natural, therefore enquiry is impiety if not blasphemy, thus reality becomes incomprehensible.[274] Freedom of conscience is not admitted; indeed there is no word in Arabic for conscience.[275] These are the reasons why Islam has contributed almost nothing to human welfare and culture during the last thousand years. These are serious charges to be laid against any culture, and I invite contradiction with examples and rational argument.

*A naturalist's perspective*

Naturalists can be irritating by saying, 'Never mind what people tell you, what *actually* happened?' The result of Muhammad's life was that, having mated with a large number of women, he presumably fathered a proportionately large number of children. However, only six are recorded, and we would surely know about any others. Since most of them and their children died prematurely of natural causes,[276] Muhammad may have carried some genetic defect. If this is the case, it helps explain why he was so sexually active – perhaps he felt a need to compensate for some faintly perceived inadequacy. This idea could also throw light on why he diverted his energies to acquiring a material empire and propagating a creed. Though these two activities are mutually dependent, to tease them apart helps understanding, and this I do in the next chapter.

Muhammad's wealth and ideas spread rapidly during his life, and even faster soon after it. Within a few years of his death, the dynasty he founded controlled a significant proportion of the world's land area and material goods. Thus, as measured by material goods and ideas, Muhammad was successful, even if he was less so genetically. However, in 2015 CE (1436–1437 AH), Muslims are one of the fastest-breeding populations, and this is why I give them a chapter in this book. Also, by an accident of geology (or the munificence of Allah – certainly not by their mutual cooperation or ingenuity), they control much of the world's oil reserves, and their faith is held by about a quarter of living people.[277] So, Muhammad's behavioural descendants are enormously successful. How and why has this come about?[65]

Muhammad was born into an impoverished land. It had been rich in plants and animals but ploughing and overgrazing had destroyed its soils and reduced its rainfall.[66] He was also born into a community which lacked political institutions that could have given the population a secure future. This uncertainty discouraged people from cooperating with each other.

[65] The long answer to this question is the subject matter of this whole book.
[66] See Chapter 10 on *Ecology & Conurbations* for an explanation of this.

Behaviourally, they were locked into a system in which family, clan and tribe were (in descending order of shared genes) more important grounds for allegiance than rationality. They robbed and killed their neighbours and travellers rather than traded with them. Thus they remained materially, socially and politically poor, though two relatively wealthy empires were neighbours.

In effect, their society was vertically stratified: children were born into a clan and it was socially unacceptable for them to change their loyalties to another clan or tribe as they grew up. They also felt obliged to share the fruits of their hard work or enterprise with their family, again at rates that were proportional to shared genes. This behaviour springs from deep in our evolutionary past and is petrified by the fatalism of faith in an absolute god. Such fatalism insists that hard work and flexible thinking are merely gifts bestowed by a god, in this case Allah, on favoured individuals. This mental furniture contrasts with that of societies in which faith is less absolute and family obligations are less severe. In such societies, individuals are freer to gain wealth, and to change their social and political allegiances. In accepting that individuals move through the layers of society, in effect, it becomes horizontally stratified.[278]

Growing up in a vertically stratified society, Muhammad perceived that robbery and war were the most effective ways of gaining enough wealth to allow him access to the many women he desired. In the event, he began with marriage and trading – with Khadija. It may well have been that he had national, if not global, ambitions as a young man – certainly he set about it in a systematic and effective way. Observing the natural superstitions of his polytheistic compatriots, and that the Jews were convinced that they were their god's chosen people, he realised that success depended on a supreme god-of-gods. The Christians' muddled three-in-one, with its foothold on Earth would clearly not do.

For five years Muhammad thought through his strategy and tactics, often consulting Khadija. To be supreme, his god could not be accessed directly by anyone, therefore he needed an intermediary, and Gabriel of the Torah was suitable. His god should not have an earthly presence, as the Christians' god had;

therefore he, Muhammad, would be the messenger. I
he persuaded his followers to regard him as the apos
described in the Torah. Later, he spoke of himself as
the prophets, and Islam as the final maturation of monotheism, superseding Judaism and Christianity.[279] He chose 'Allah', from his tribe's deity, as the name for his god, and thus preserved some continuity.

Muhammad's revelations are presented to his followers as coming from Allah via Gabriel. It is more realistic to think that the revelations were dreams or wishful thinking, rather than involving the truly supernatural in which the laws of physics are suspended. Whatever his revelations, they were so effective in achieving simple biological aims that it is likely their subject matter was deliberately planned.

His followers saw their lives as similar to Muhammad's. They were mostly poor, and he presented them with an opportunity to gain material wealth and access to women. He legitimised murder, robbery, extortion, rape and breaking of treaties, so long as the followers obeyed him. He declared that these commands were the will of Allah, and so not open to question or discussion. He so convinced his followers of this that they accepted him taking a fifth of all that they stole, and also his having first choice of captured women. His instructions to them were to kill men, to mate with women and bring up children as Muslims; this is a clear biological tactic to increase the number of Muslims.

In the 21st Century CE (15th Century AH), the ecological and economic situation in the world is similar to that of 1st Century AH Arabia. In spite of vast oil wealth in some territories, ordinary people in lands controlled by Muslims are generally poorer than their counterparts in developed countries. These poorer people have immediate visual access, via social media, TV and so on, to that material wealth.[280] Envy is a powerful motive, and it is exacerbated by the difference between rich and poor within those Muslim lands.[281]

Because they are less willing to doubt, criticise and think rationally, strict Muslims are unable to compete with the skills of people in developed nations.[67] For them, the easier and more socially acceptable course is to follow the pattern of Muhammad's life. The logic runs like this: since Muhammad was successful in accumulating wealth, having access to women and spreading his ideas, it is sensible to do as he did in the way he did it. Certainly this course of action is acceptable to other Muslims. Indeed, it is what they are taught to do in Muslim schools. This explains why Muslims do not directly condemn atrocities or sexual exploitation committed by Muslims against kaffirs, but only mouth generalities.

As I explained at the end of Chapter 5, human sexual behaviour is at a stage in evolution between the promiscuity of chimps and the monogamy of gibbons, though descended from neither. Sexual competition between males is a major source of social tension and conflict where primates live in groups. Human evolutionary success depends on cooperation between sexually active males in close proximity. Since men and women are normally born in equal numbers, polygamy deprives some men of heterosexual partners, and so creates tension and undermines cooperation between them. This non-cooperation is controlled by submission to a powerful leader or deity who will allow some access to females by those males that cooperate. In giving captured women to his soldiers, after he had taken first pick, Muhammad was doing exactly what alpha chimps do, and for precisely the same motives.

Monogamy is a different solution to the same problem of male rivalry. It differs from permitted promiscuity in that it relies on trust, and trust allows innovation. Trust allows a higher density of men in any one place, and this, as we have seen, has been heavily selected for in our recent evolution. Monogamy is by no means perfected in any society, even serial monogamy deprives other men less than does polygamy.

**********

---

[67] See also Mill (1859: p. 75) on the need for diversity of opinion, if social (or any other) progress is to be achieved.

Islam's publicly stated ambition is to force the whole world to become Muslim and accept Sharia law.[282] They are unwilling or unable to do this by rational argument, so a minority of them resort to terrorism and open warfare. A more sophisticated majority of Muslims have perceived that kaffirs of the richer nations generally regard human life as sacred. This encourages Muslims to achieve their reproductive aims directly; which is to say that they breed as fast as they can and rely on the kaffir's concept of morality to accept into their countries the excess Muslim population. The present flood of migrants entering Europe and elsewhere are mostly Muslims, and a significant number of them are jihadists from so-called Islamic State.[68]

The rising proportion of Muslims in European countries represents a significant political lobby. Fear of losing their votes is one reason why European politicians will not confront Islam's stated aims. More significant is that Muslims are committed to destroying the liberal regimes in which they have taken refuge, and converting them into the sort of theocratic dictatorship that Muhammad directed,[283] and which has led directly to their present poverty. In effect, they are a fifth column of Islam.[284]

The concept of the nation state whose people can freely choose by whom they are governed is anathema to Islam. A Muslim who lives in, for example, England is not an Englishman who happens to be a Muslim, but a Muslim who happens to live in England.[285] This is why the first question at the beginning of this chapter is crucial. Anyone who answers it with 'yes' effectively seeks to hand control of the land they live in to Mecca. That control is 'top down', which is to say that political policies are devised and decision made by a religious dictator, or perhaps an oligarchy, and the results imposed upon the mass of people. The system Islam seeks to replace, at least in countries based on English Common Law, allows and

---

[68] *Spectator*, 20 June 2015, p. 17; BBC News Radio 4, 6 AM, 17 May 2015. Concern about migrants continued to make headlines throughout the summer of 2015. I predict that each year the problem will multiply until the root cause, overpopulation, is faced and at least discussed. So far, it has not been, because the subject is taboo.

encourages all manner of thoughts, discussion and assemblage of people,[286] and expects those ideas to filter upwards, if necessary through courts, to parliament itself where they may be engrossed into Common Law.[287]

The difference between these two systems is crucial to the future: Islam imposes conformity, uniformity and absolutism, and it will not tolerate opposition. It sees democracy as a step towards a narrow hegemony. 'One man (literally), one vote, once,' is their creed.[288] On the other hand, science and Common Law encourage diversity, doubt, questioning and criticism. They thrive on opposition and peer review, and they continually seek correction of old ideas and suggestions for new ones. Islam will not change in the face of changing circumstances, Common Law can and does. And it is why, in the long term, Islam condemns itself to extinction, whereas science and rationality, as expressed in Common Law and peer review, will survive. Fourteen billion years of evolution have taught us that, and Islam denies it. That is why Islam is a special case.

Such denial would be fine as a private conviction because everyone is entitled, in a non-totalitarian world, to believe what they like – provided that belief does not stop others from believing what they like. But Islam is a totalitarian political movement thinly disguised as a religion with a powerful proselytising drive written into its fundamental tenets. This, combined with Muslims' determination to dominate by breeding (open warfare, terrorism and convincing argument having failed), will not only ensure Islam's own eventual demise, but also destroy the rational kaffir world.[289] That is why Islam must be opposed. So how do we kaffirs do this?

First, because it is the basis of success, rational discussion must be offered to all people, hoping to persuade them to cooperate in safeguarding future generations. If rationality fails, and opponents of it resort to irrational reliance on sacred texts or their animal feelings, it makes more sense to counter their oppressive activities with other means. I give one example.

Mere force does not inhibit an Islamic martyr because they know for certain that, if they die while killing kaffirs, they will go to heaven and live forever in conditions of perfect pleasure.[290] They also know that they will not, if they are contaminated by, for example, pigs. These certainties were used in John Masters's novel *Bhowani Junction*. In the book, which is set in India, Colonel Savage orders his Ghurkhas to urinate on high-caste protesters who are lying on a railway track. They flee in terror, knowing that they would lose caste if they become contaminated. The episode was sanitised to some extent in the film version with Untouchables bringing buckets of pig manure to throw on the protesters.[291] Perhaps, if something similar could be devised for the bodies of Islamic martyrs, might it reduce recruitment of their replacements? This book *is* a biological commentary.

**********

Wearing distinctive clothes is almost always a public statement about the wearer's beliefs, and should certainly not be forbidden in a free society. Nor should that freedom be separated from the duties and responsibilities that are indissolubly bound to it. Thus, if a woman is free to wear a headscarf as a symbol of her Muslim faith, she also has the duty to accept the reaction of other people to it. If that reaction is approval by co-religionists, then she will feel a bond of common interest and ambition. If, on the other hand, people she meets express disapproval, then she cannot object. There being no right of freedom from offence, we are all entitled to make our views plain, though not in a way that might incite disorder. If she argues that her headscarf is intended to curtail lascivious thoughts in men, then perhaps she should seek a society in which men either do not have them or lack the freedom to express them. If she chooses to live in a kaffir country (assuming she has a choice), she and her menfolk must accept that those around her understand that the ambition of Islam is to ultimately destroy their way of life, and the headscarf can be seen to be a declaration of that intent.

Women who cover their faces are hiding the one feature that marks each one of us out as an individual, thereby signalling that they are merely objects, which certainly Muhammad took them for. This, of course, is the aim of a criminal who does the same, and equally applies to soldiers who wear masks, because they do not want to be recognised when they return to civilian life, which is their normal situation.

In sum, it is insufficient to attack Islam by pointing out its deficiencies and dangers. More important is to explain the strengths of the systems it seeks to replace. And the most vital category of these includes individual freedom of conscience, assembly and expression. A close second are the institutions that protect democracy, such as a secular and freely elected parliament and well-separated legislature, executive, judiciary and police. All of these are the products of organic evolution that is behavioural rather than genetic. These institutions all found their, by no means perfect, development under Common Law; however, the essential separation between institutions of state are obliterated when bound to Islam.

**********

In this chapter, I run the risk of being denounced by *fatwa*, and possibly my life threatened. In the true spirit of an Islamic martyr, I welcome such a fate, not because I seek eternity, with or without energetic maidens of suitable age, but for two other reasons. First, I was born in 1938 and my life is nearly done anyway. Second, because anyone making such an edict or carrying it out would give this book the publicity and immortality that I think it needs.

# Chapter 9

## GENES & 'THENES', IDENES & INDIVIDUALS

So what are we really trying to do with all this eating and breeding, talking and writing? We can think of a book in two ways: the paper or computer files on which the words are recorded, and the information contained in those words.[69] That information can be combined together in ways that carry larger ideas. It is easy to print more copies of the book, and doing so merely multiplies copies of the information it contains. In exactly the same way, a gene exists in two parts:[70] the information it carries and the medium in which that information is recorded. The natural medium is the sequence of nucleotides in the chromosome, as explained in Chapter 3. The same information could equally well be stored as the letters C, G, A and U, remembered in the human mind, written on paper, or as a sequence of magnetic or other irregularities on a tape or disk. This has been done for the human genome, though it does not mean to say that a human being could be reconstructed from that printed book.

When we die, our bodies decompose and the unique assemblage of genes, which we regard as the individual each of us calls 'I', disintegrates. It disintegrates but need not necessarily be lost. If we have children, much of our genetic information is passed on in our germ cells – it is potentially immortal. Each of our germ cells contains half the information in each of our body cells. To make a new individual, that half set of information has to cooperate with the half set in a germ cell

---

[69] I use the word 'information' to include 'knowledge' in this discussion.

[70] Williams (1992: p. 11) writes: 'A gene is not a DNA molecule; it is the transcribable information coded by the molecule.' Quoted by Dennett (2006: p. 349); see also Dennett (1995: pp. 348, 355) and Wilson E. (1998: p. 149).

from another person, and it does so by combining with it. This is the sexual process of reproduction. In other words, half of what ultimately makes us genetically unique is passed to each child we have. This does not mean to say that two children will carry a person's full genome because there is reassortment at each recombination. The more children we have, the more of our genetic individuality is likely to survive in the next generation, though shared among more progeny. That is our genetic legacy, but there are other forms of immortality.

Individual Pleistocene people who used tools left more offspring than individuals who did not: they were evolutionarily more successful. Likewise, groups of individuals that could live at higher density by farming left more offspring than groups that did not. The possession of tools and a food store came to be seen as an asset by critical partner-seekers. When the owner of these things died, they represented something about him or her. Maybe this is why funerary rites reflect, if not display, dead people's material wealth, though it may not go so far as to bury with them many of their earthly possessions, as happened in ancient Egypt.

It is cumbersome to think of stored food, tools, cattle, cars, houses, land, jewels or money as 'things', so I give the name *thenes* (to fit with 'genes') to all material things that can be passed from one generation to the next. Thenes, not genes, are now the common currency of success in the Western world. We quietly pride ourselves on our wealth, rather than the number of our children; though, in poorer parts of the world, children remain the chief asset. Thenes are a person's material wealth, however it may be measured.

Just as a book or a gene consists of two parts, so also can we think of thenes as being both the actual objects within that category and also as having a value. That value need not necessarily be expressed in money: a great work of art is worth more than so many pounds or dollars, however much the mass of people want to hear a figure mentioned. Prestige in ownership, aesthetic satisfaction and scholarly achievement in assembling a collection are examples of other values.

Parents propagate genes by producing children, and they do so more efficiently by accumulating thenes, but grandparents throw a different light on inheritance. Fathers have spent much of their lives displaying their hunting (working) skills to each other in hierarchical struggles. They may also have spent time defending, or preparing to defend, the family's territory by repairing the house or displaying a fine garden, or the group's territory by military service, or just by paying taxes.

Once the mother of their children is past childbearing, the best interests of a man's genes are served by encouraging him to find another, younger woman. In the past, those who did so left more children than those who did not. That is why men can go on fathering children practically all their lives. Men generally die younger than women because, since ancestral men were more likely than women to be killed while hunting or in fights with other men, natural selection did not favour investing much physiological effort in repairing their senile tissue.

Women are different: they not only live longer than men but also become sterile in their forties or fifties when they reach the menopause. The menopause has evolved because it was advantageous to the genes that caused it. Giving birth is risky to the mother, especially in humans where the baby's relatively huge head has to be forced through a pelvic opening that is specially narrowed to allow an upright stance. But an extra pair of hands in the cave is a great advantage when there are young children about. So also is the store of remembered experience: which plants are nutritious in which season, which are poisonous or have healing properties, and so on. Thus, if a woman stops running the risk of having children herself and helps to look after the copies of her genes in her daughter, she is more likely to leave more *grand*children than a similarly aged woman who goes on having babies. The risk is multiplied because, if a woman died in childbirth, it was very likely that the younger of her previous children would die with her, were it not for grandmother.

In addition, the accumulated experience and judgement of previous generations is particularly valuable because, if a family lost all its material possessions, its members could build them up again with the skills that had been learned and stored in older people's memories. Neanderthal women had a life expectancy of about 40 years, but, over a short period of evolutionary time, Modern Women's life expectancy increased to 60.[292] As parasites evolve to attack tissues whose genetic formulation had been around for a long time, so the physiological cost to the host of repairing those tissues increased. For average lifespans to suddenly rise by 50%, there must have been a powerful selective advantage at work to outweigh the costs; and that advantage was surely the survival value of being able to pass on knowledge. In preliterate societies, grandparents were the means of storing and handing on this knowledge. This is why we evolved the menopause, and why women live longer than men. Evolutionary success is now measured not just by more offspring, but by future generations. For the first discernible time in vertebrate evolution, an animal species is projecting its genetic interest beyond its children.

**********

In Chapter 6 we saw how our species has been through two genetic bottlenecks that severely reduced our genetic diversity. In our journey from southern Africa to the rest of the world, we passed through a vast range of environments. Our genetic evolution was so slow that we had to adapt in other ways. A highly adaptable body, collective cooperation and our huge brain with its capacity to access all levels of intelligence made the adaptation simple: we evolved behaviourally.

This is clearer if we look at one imaginary example of a behavioural idea. A group of hunters, armed with sharpened sticks, find that they can injure a land animal sufficiently for them to be able to catch it. But, when they attack a seal with a sharpened stick attached to a cord, the stick pulls out and the injured seal is lost. In one imaginary hunt the spear did not pull out and the hunter landed the seal. He noticed that he had failed to sharpen his spear correctly and had left a peel of hard wood projecting backwards near the tip and this had snagged in the seal's flesh. The idea of a barb was born. It was copied many times by fellow hunters, varied, some with improvements; the best modifications were selected and the poorer versions discarded. The result of this behavioural evolution was that a band of people could occupy an Arctic habitat.

Notice that it was the barb that was evolving, not the hunter who made the spear, which is to say that selection was acting on the ideas in his mind. Since ideas can multiply, vary and be selected in seconds, evolution can proceed almost instantaneously. Compare this example with a predator that had to evolve backward-pointing teeth to catch slippery seals. In competition between the two methods, behavioural evolution wins every time. In this way we have adapted to such diverse environments as Arctic seas and rainforests, subdeserts and mountain plateaux. This is how we have overcome the limitations of the genetic uniformity that evolutionary bottlenecks had bequeathed us. In sum, we are a biospecies that has begun to 'etho-speciate'.

Richard Dawkins is a household name today. His field is evolutionary biology, but his excellent research was later overshadowed by his mastery of explanation. Between 1995 and 2008, he was Simonyi Professor for the Public Understanding of Science at Oxford. Dawkins first suggested that ideas could be units of evolutionary selection in his first and most famous book, *The Selfish Gene* (1976). Genes can multiply, vary and be selected – and this is evolution. In a way that is exactly analogous to the evolution of genes, ideas can multiply, vary and be selected. Dawkins called such ideas *memes*, and the word is now sanctified by inclusion in the Oxford English Dictionary. Though Dawkins coined the word meme, the concept was described in 1971 and called an *idene*.[71]

Just as the word 'gene' can mean two distinct things, so also can 'meme'. It can mean the idea itself, or it can mean the

---

[71] Dawkins, R. (1989: p. 192); Dawkins coined the word *meme*, but the concept was used by H. Hoagland in 'Reflections on the purpose of life', published in *Zygon*: *Journal of Religion and Science*, 1971, pp. 28–38: 'Henry A. Murray has coined the term "idene" in relation to social evolution analogous to gene in biological evolution.' 'Culture gene' was an early phrase for meme in Dawkins's sense (Cartwright, 2000: p. 311). Flannery (2010: p. 18) challenged this.

medium in which that idea is recorded. I shall use the word *meme* to mean the medium in which an idea is recorded, and *idene* to mean the idea that is recorded. Thus idenes are stored and transmitted as memes.

Table 9.1 shows a summary of the dual aspect of the three classes of entity.

| | |
|---|---|
| **Gene** | |
| Genetic information | Medium in which that information is stored and transmitted, for example, a nucleotide sequence, printed text or a computer programme |
| **Thene** | |
| Wealth that possessions represent: money, social status, aesthetic thrill, and so on[293] | Possessions in which wealth is stored and by which it is transferred: 'things' |
| **Idea** | |
| Idea itself (idene) | Medium in which the idene is stored and transmitted, for example, neuron constellation,[72] musical score, printed word, film, computer program (meme) |

**Table 9.1** Gene, thene and idea

These divisions are getting near the Aristotelian idea that every proposition must consist of two parts and, corresponding to these two parts, reality itself must be divided into substance and attribute, the latter being inherent in the former.[294]

**********

[72] Delius in Dawkins, Halliday & Dawkins (1991: p. 83) shows a photograph of 'a meme as a constellation of activated neuronal synapses lodged somewhere in the brain of an individual.' Richard Dawkins also supports this by: 'memes must be actual brain structures, actual patterns of neuronal wiring-up' (Dawkins 1989: p. 323).

John Stuart Mill (1806–1873) was born and lived in London, though of Scottish descent. He published extensively on a wide range of philosophical subjects, and I shall look at one of his propositions. It was that, for him, the test for logic, as for everything else, is the ordinary world as we ordinarily experience it. We, in the 21st Century, experience a different world from Mill's. It is different in two ways: obviously the world is more crowded and urbanised, and the information we have available to interpret what we experience is vastly different.

Consider the syllogism:

'All men are mortal.
Socrates is a man.
Socrates is mortal.'

Mill quoted this as an example of deductive reasoning but I shall use it to illustrate another point. By 'a man', we mean a recognisable individual such as Socrates, complete with his warts, scars, reproductive organs, gut flora, parasites, memories, wisdom and personality. By 'mortal' we mean that all these components and attributes will disintegrate at death. Recall that 'disintegrate' does not necessarily mean destruction or extinction; it implies dispersal, and the different components will disintegrate and disperse with different destinies. The proposition, 'All men are mortal' is based on an outdated concept of what we now understand it really is to be an individual. First, we can leave aside medieval superstitions that burden us with souls – a concept articulated by Aristotle, and scarcely questioned since.

The second suite of ideas that influences our thinking is more complicated. For example, in December 2013, I met a man called Malcolm to whom I had taught biology nearly 50 years before. I remember him as a small 14 year old. Now he is taller and heavier, and with white hair, but his face is still recognisably of the person I knew long ago. Is he the same individual? Yes, because he remembers me, and I remember him. But there is a difficulty. Living tissues are continuously changing: cells die and their components are shed or excreted, and new tissues constructed from food.[295] This means that, apart from some densely ossified bone, and natural dilution, nothing material of the Malcolm I knew as a boy is the same as that of the man I met recently. His and every other living individual creature's true condition can be thought of as analogous to a standing ripple on a river (but not to a spreading ripple on still water.)

Though materials flow through a river ripple, its individual shape and behaviour remain recognisable. They are determined by the speed of water, its air content, the form of the substrate and other physical parameters that can be measured and recorded – they are information, and can be converted into knowledge. In the same way, materials flow through our bodies to replace our substance. How those materials are assembled and how they behave are controlled by different and vastly more complicated information. Much of this information is carried as a digital code in our genes. Another kind of information has impinged on us in the form of external influences that have, in many cases literally, shaped us. Though there is significant functional overlap between genes, structure and behaviour, it helps understanding if we think of this information as two main kinds: genetic and behavioural.

Genetic information is of three kinds. We inherited half of one kind from each of our parents, and it was fixed at conception. The second, mitochondrial DNA, we inherited from our mothers only. We acquire the third kind after birth, and it is carried in the kilogram or two of microbes that inhabit our individual skins and guts.[296] Among these microbes are species that manufacture vitamins, help digest our food, cause ailments or defend us against such pathogens. Indeed, without the benign sorts we would quickly die. They probably contribute to our individual smell, which is largely subliminal in humans – something my dogs do not understand. We readily exchange these microbes with other people, the pathogens appearing as diseases.

Behaviour is less tidily classified: some of it is controlled directly from inherited genes, and it blurs into reflexes and physiology, but the most conspicuous category is learned.[297] Malcolm's experience of the world is vastly greater than that of the bright boy I knew. He has read more books, argued with a wider circle of acquaintances and been uplifted or hurt by intimates. He has thought about his experiences and then modified his behaviour, consciously and unconsciously. In the sense I use it here, his behaviour includes his knowledge, habits, mannerisms, cast of mind, opinions and prejudices. They all contribute to making him the recognisable person that he is.

Everyone he meets will be influenced by the product of that experience, and he will pass on the ideas that he received in those transactions to the next person he interacts with, altering *their* accumulated experience, and so on. Like the information in his genes, Malcolm's ideas too can live forever, even if, like them too, in greatly diluted form. So, making allowance for the ageing process, what is it that I recognise in my former pupil? It is the information he carries within him, and not the atoms I see before me. If he had experienced a disfiguring accident or a behavioural catastrophe, such as a personal tragedy or Alzheimer's disease, which changed his appearance or behaviour beyond what I expected, then I might say that I did not recognise him.

Though physical appearance is the immediate stimulator of recognition when two people meet, our patterns of behaviour are what characterise each of us more surely as an individual. Those patterns can be recorded in nervous pathways, on film, and as sound recordings, images or even as bottled smells: they are our idenes, and they too are distinct from the medium in which they are recorded. Though the recording medium may perish at our individual deaths, we will have already passed on our idenes to every person with whom we interacted – they, like our germ-line genes, are potentially immortal.

We can think of each individual living creature as a period of interactions between materials and information. These two components can be seen as two streams flowing through time. Materials pass through our bodies during our lives, replacing tissues and eventually being excreted. The information that controls how these materials are assembled and behave is also a stream whose composition is continually changing. The information flowing through each of us is the tiny trickle that interests us individually between conception and death. That trickle is part of a wider and deeper river of information that is not only shared, at least partly, by all living things but extends down into the nature of chemistry and subatomic physics. I have long looked for evidence that there is a real break, as distinct from a gap in our understanding, in the sequence of physico-chemical events between the Big Bang, if such an event ever happened, and my consciousness as I write these lines. So far, I have failed. But then how can one prove an absence?

From these thoughts it appears that what each one of us calls 'I' is no more than a temporary vehicle on which natural selection acts, thus adjusting the composition of the whole river of genetic and behavioural information that flows through time. Though the idea of 'I' is the centre of our individual existence, in reality it is as inseparable from all life as is a grain of dust from the roads of the Earth.

I think that John Stuart Mill had only the physical body in mind in his syllogism. This is a pity because Socrates's ideas are more important than his body – including his testes. In the synthesis I outlined earlier, Socrates is immortal; that is why we are still talking and writing about him.

In the same way, each one of us is immortal: our genes, thenes and idenes can be handed on to succeeding generations. Even if we have no children, die in poverty and fail to generate a single good idea, we can still live a useful life by helping others to preserve and propagate the genes and idenes we share with them. This is how the individual appears to behave morally, even though their genes and idenes are behaving selfishly.

**********

Our modern world is dominated by thenes, and their values are commonly expressed as money. Money has profoundly influenced the way our cultures have evolved. A Martian visiting the Earth for the first time, and having seen at once that our main purpose is to breed, would surely take the view that money was the principal manifestation of a religion, if only as the repository of faith; faith that was badly shaken by the economic troubles which began in 2008, and are still getting worse (as measured by the growth of total debt) in 2015. Money is, perhaps, the most powerful distorter of thinking rationally, especially about such a capital sum as London. We should look at it more closely. But, before we can do that maturely, we need to understand how natural systems interact as the materials cycle within them and energy flows through them. This study is called *ecology*. The concepts of the ecosystem and the mechanism of evolution are the two fundamental mechanisms that have operated throughout Earth's existence.

# Chapter 10

## ECOLOGY & CONURBATIONS[73]

Only when we have appreciated the ideas in this chapter will we truly understand the contradiction in the phrase 'sustainable growth'. Building a new London every five weeks is growth but few can think it is sustainable for long, let alone indefinitely.

London is an *ecosystem* though rather a complicated one; indeed it is too complicated to unpick all its components to understand how it works, so we will start with something simpler. An ecosystem is any part of the living world we choose to think about. 'Part' includes time as well as space: are we going to think about the space we define for a day, a few weeks, for a year, or more? We can make a simple ecosystem by planting a seed in soil in a flowerpot.

It will help to understand the simple ecosystem we have made if we divide it into three components: (1) **living things**, that is, the seed, any centipedes, earthworms in the soil or slugs that might eat the seedling, and other animals that might eat them or each other, and fungi and bacteria,; (2) **soil**, a complex system of mineral particles, dead organic matter, air and water; and (3) **physical parts** of, and influences on, the ecosystem. Physical parts comprise: *air*, including pressure, draughts or wind speed; *water* that we add to the pot and in the air as humidity; *chemicals*, such as salt, plant nutrients, acids, man-made poisons, and so on; light intensity and duration; and *heat/temperature*, both average and fluctuations. Living things

---

[73] Much of this chapter is elementary school biology. Its content and style evolved over the 18 years that I taught 11–19 year-olds, so the sources are lost. However, many school biology textbooks contain this information. See also, for example, Andrewartha & Birch (1984), Billings (1964), Cherrett & Bradshaw (1989), Dickinson & Murphy (1998), Elton (1953), Greig-Smith (1983), Kormondy (1969), Lambert (1967) and Lomborg (2001), who makes many basic mistakes on which to found his scepticism. See also Neal (1953), Odum (1963) and Solomon (1969).

and soil are present in fairly constant quantities in our artificial ecosystem, but water, air, light and heat are continually being imported and exported.

We can make a *closed* ecosystem by planting an African violet in peat in a large glass storage jar, lightly watering it, closing the lid and leaving it in indirect light. (Sunlight will kill the plant.) It will survive well for several weeks, though eventually the system will break down when diseases get the upper hand. It is closed because no materials can get in or out, yet the African violet goes on living. Therefore, materials must be circulating in the system.

Water evaporates from the leaves of the plant and the peat surface, condenses on the glass and runs back into the peat. Bacteria and fungi in the peat break it down releasing carbon dioxide into the air. The African violet takes in the carbon dioxide and turns it into organic material, which will add to the peat when the plant dies. The plant also takes up mineral salts dissolved in the water that percolates through the peat. When the fungi and bacteria die, they too add to the peat. Disease-causing parasites will feed directly on the violet, and eventually kill it.

We can present this information as a diagram to make it clearer. The arrows show the direction in which materials are circulating (Figure 10.1).

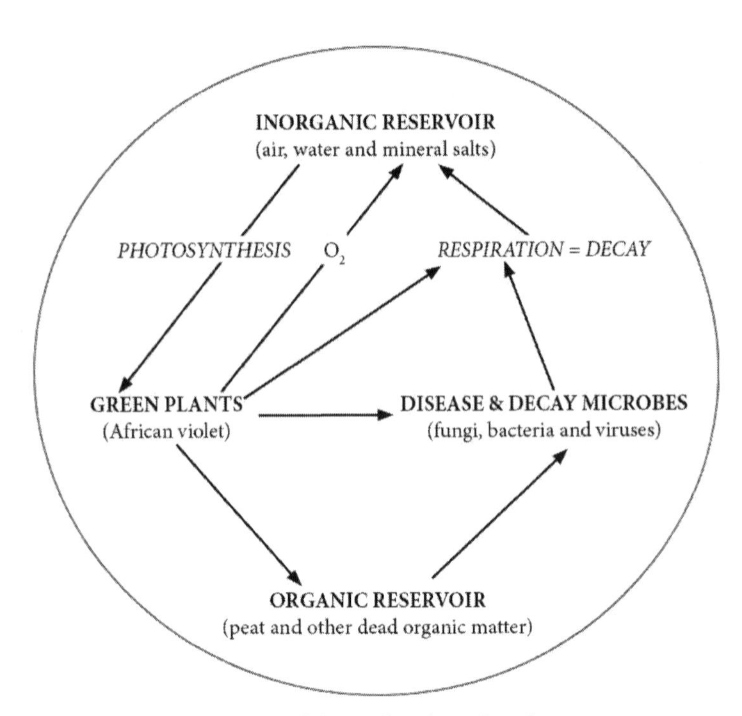


**Figure 10.1** Materials cycling in a closed ecosystem

To drive this circulation of materials, there has to be a through-flow of energy. The whole of our ecosystem is at about the same temperature. There is certainly not enough of a temperature gradient for heat to flow from one place to another, which is the only way in which it can do work, so heat cannot drive the processes. The only alternative energy source is light. Green plants, such as the African violet in our jar, can turn light energy into chemical energy by the process of photosynthesis. Enzymes in the African violet now release that chemical energy it has stored to do the physiological work necessary for the plant's life processes. The chemical energy in the organic molecules eventually turns into heat that is lost by radiation and convection. Again, this is clearer in a diagram. (Figure 10.2)

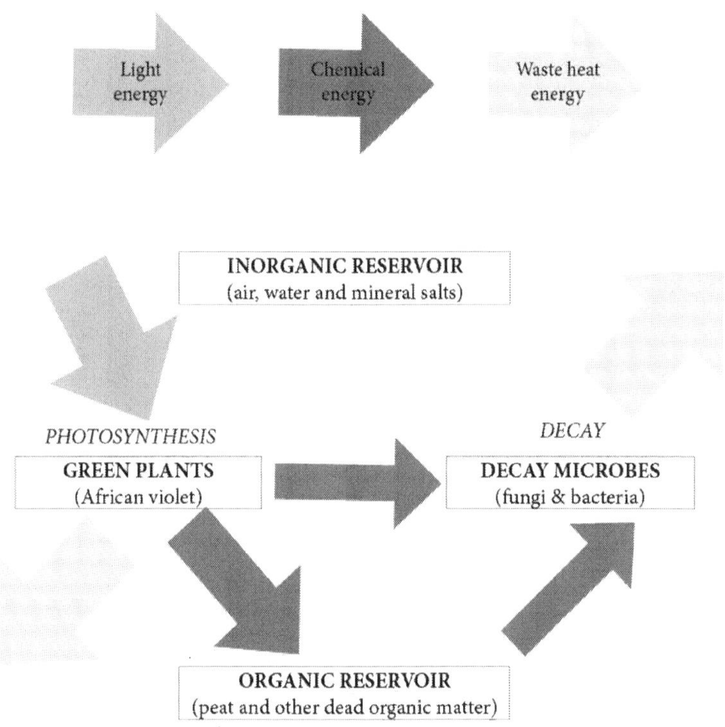


**Figure 10.2** Energy flowing through an ecosystem.

While the plant is growing, the ecosystem takes in more energy as light than it gives out as waste heat; this surplus of light energy is stored in the plant's living tissues as chemical energy. When the plant dies, its body adds to the organic reservoir of peat, carrying its store of chemical energy with it.

Taken together, Figures 10.1 and 10.2 show that materials circulate in an ecosystem, and energy flows through it to drive this circulation. Like our storage jar, the Earth is a closed ecosystem. Apart from a few tonnes of meteoric dust and rocketry, no materials are gained or lost from the planet. On the other hand, sunlight energy pours in by day, and radiant heat pours out by night. The balance between the two determines global warming or cooling.[298]

If we put our closed ecosystem into the dark, the plant will certainly die, but the decay microbes will go on living until they have used up the organic reservoir of energy. If we mentally substitute oil for light, the parallel between this situation and the present human predicament is obvious. Is it stretching the metaphor too far to suggest that substituting money for light is having a similar effect on economies? Perhaps we can imagine light energy as equivalent to creating new money, and waste heat as monetary inflation. Chapter 11 explores this idea.

Now we need to add some complications to make our model more like the ecosystems found on Earth. In Figure 10.3 the arrows represent both energy and materials.

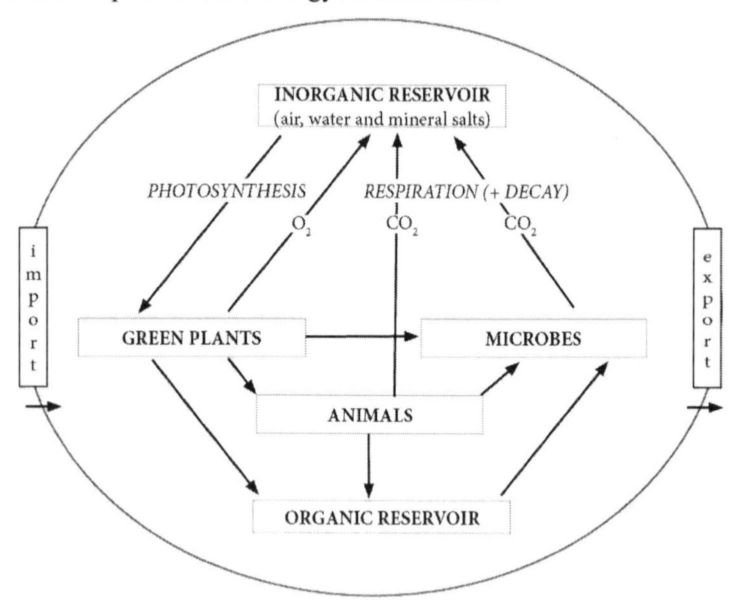


**Figure 10.3** Energy and materials flowing in an ecosystem.

Next we can divide the animals into two classes: plant-eaters (herbivores) and animal-eaters (carnivores). The relevance of this is summed up in the T-shirt that read, 'Save trees – eat a beaver'. In the same way, we can think of wolves as helping plants when they kill or alter the behaviour of herbivorous animals.[74] Many animals, like us, eat both plants and flesh, and we need to imaginarily divide their feeding between the two classes.

So far we have looked at what kinds of materials and energy have flowed in the ecosystem; now we need to consider the *quantities* of material in each of the five categories: (1) green plant; (2) herbivorous and carnivorous animals; (3) decay-causing microbes; (4) inorganic reservoir; and (5) organic reservoir. It is difficult to quantify the last three, and we will have to make some hefty assumptions about the others. Anyway, what do we mean by 'quantity'? Clearly it is unsatisfactory to merely count individual plants and animals because they differ in size. More sensible is to estimate the total mass of each category by weighing it. Some organisms are wetter than others – it makes little sense to compare jellyfish with limpets – so comparing dry weight (biomass) gives a better measure, though I do not advocate drying live animals.

From many measurements in a variety of different ecosystems, we can work out that transferring organic matter from plant to herbivore to carnivore loses 90% of the material at each exchange. The lost material is used up to release chemical energy needed for the day-to-day life processes of the animal that is consuming it, and only 10% is stored in that animal's body. This is why carnivores are so rare compared with herbivores, which are rare compared with plants. The difference between 1 biomass unit of carnivore, 10 units of herbivore and 100 units of green plants is even more striking if we represent these figures pictorially as a *pyramid of biomass* (Figure 10.4).

**Figure 10.4** A pyramid of biomass.

[74] For example: wolves let beavers recolonise Yellowstone National Park by altering elk browsing behaviour and so allowing riverine vegetation to increase (Douglas Chadwick, *Wolf Wars*, National Geographic Magazine, March 2010).

**********

This general principle relies on all the chemicals being processed (metabolised) by the plants and animals in our ecosystem. What happens to chemicals that they cannot metabolise? Fat-soluble types often accumulate, and here the idea of a food chain is important. Let us suppose we use a fat-soluble insecticide that is lethal to birds at a concentration of 70 parts per million (ppm). We dress our metric ton of seeds (we are farmers) with 1 gram (g) of insecticide. This is a concentration of 1 ppm and safely within the lethal limit to birds but fatal to insects. All the seeds are eaten by 100 kilogram (kg) of sparrows (we are incompetent farmers), which digest the organic matter and store one-tenth of it in their bodies. However, they cannot metabolise the 1 g of insecticide, so all of it remains in their bodies. The concentration of insecticide in the sparrows is now 10 ppm – still not lethal. All the sparrows are eaten by 10 kg of sparrowhawks (they are incompetent sparrows). The sparrow flesh is digested and one-tenth of it is converted into sparrowhawk. Again, the 1 g of insecticide that was in the sparrows is now in the 10 kg of sparrowhawks, so its concentration is now 100 ppm. Dead sparrowhawks. (Table 10.1).

| Living organism | Total weight of organism | Total weight of insecticide | Concentration of insecticide |
|---|---|---|---|
| Sparrowhawks | 10 kg | 1 g | 100 ppm |
| Sparrows | 100 kg | 1 g | 10 ppm |
| Seed | 1000 kg | 1 g | 1 ppm |

**Table 10.1** The concentration of a material that cannot be metabolised increases as it passes through the food chain.

**********

Understanding the concept of the ecosystem is fundamental to every activity in the biosphere. It is the balance sheet that should be applied to *any* system, especially businesses and cities. Business or civic managers may run their organisations efficiently: they know how raw materials, storage facilities, manufacturing stages, waste processing, import and export are interrelated, and how energy and money flow through their systems. These components are under their control. But they do not have control of the original sources of the energy and raw materials they require (the imports into their systems), nor do they ultimately control what happens to their products and waste (the exports from their systems). How thoroughly have they planned for the time when imports fail or exports or waste accumulate?[75]

Every city can, and in my opinion should, be thought of as an ecosystem. Apart from a negligible number of trees, lawns, backyard vegetables and solar panels, a city does not trap the energy it needs from sunlight. It does not regenerate the oxygen it consumes, nor does it produce its own food or raw materials, nor does it recycle its own waste. It depends on the surrounding countryside, or farther afield, to do these things for it. A city is little more than a system of import and export. The wind brings in oxygen and carries away carbon dioxide and pollutants, and rivers bring water and carry away sewage often without cost of transport.[299] Everything else, all the energy in the form of gas, oil or electricity, all the food, all the consumable products not manufactured within the city, all the raw materials of its industry must be brought in by trucks, trains, planes or ships at great energy cost. Liquid waste must be drained away and treated in sewage works, solid waste must be carried away and dumped, also at great cost. Think of London …

Medieval cities were limited in size by the difficulty of transporting food and disposing of sewage. Embanking the River Thames and building sewers in them reduced a major impediment to London's growth. Piped water allowed the next advance in size; then trains and railways, and lorries and roads. I wonder what the limit to a city's size is. Currently (2015) the conurbation around Tokyo holds the unenviable record at nearly 40 million inhabitants.

If transport systems break down in a modern city, there is serious trouble, as strikes have indeed shown. In fact, transport

---

[75] Economies are subsets of, and ultimately dependent on, ecosystems. See Daley, Cobb & Cobb (1994), quoted by Polunin (1998: p. 59).

has become more and more efficient as oil extraction and processing (*not* production) has increased. Therefore, cities have been able to import their needs from farther and farther afield. Fruits and vegetables are now neither seasonal nor local; they come from all over the world. Like ants' nests, cities spread their tentacles out into their surroundings and strip it of what they need, and dump what they have finished with.

Apart from a few pets and pests and a very few wild creatures, cities contain just one species of animal, *Homo sapiens* and our nests and toys; they are almost monocultures.[76] Like all monocultures, cities are very vulnerable to plagues; indeed they are themselves a plague. Plagues are the exponential increase in a population of living things such as Biblical frogs, lice, flies, locusts, lemmings, rats, rabbits in Australia, aphids on beans, caterpillars on cabbages or in forests, bubonic and cholera bacteria, and the Ebola virus. Plagues are commonness out of control. We are a plague.

Plagues occur in the Arctic, in the desert, in farm crops, and in crowded cities. They do not occur in rainforests, coral reefs, savannah or mixed woodland. Why not? Lions feed on zebra, wildebeest and other large grazing animals on the plains of Africa.[77] If zebra are the most common prey species in an area, the chances are that the local lions will kill and eat them more often than rarer species. When zebra become scarcer, chance will supply the lions with the new most common species. In such diverse ecosystems, there is a penalty in becoming common; commonness will never get out of control to become a plague. This is another meaning of 'the balance of nature', and

---

[76] Elton (1953: p. 17). *Oligoculture* – an ecosystem with only a few species of plant or animal living in it; *monoculture* – an ecosystem with only one species. In practice a monoculture is unlikely, though farmers try to achieve it. *Polyculture* – many species. The need to maintain biodiversity is paramount to our survival. See Wilson E. (1994: p. 359) and Cohen (1995: p. 339).

[77] Andrewartha & Birch (1984: p. 119) quote Schaller who concluded that lions have little impact on wildebeest numbers; however, Sinclair in Cherret & Bradshaw (1989: p. 222) mentions circumstantial evidence that they do. Even if not strictly accurate, the illustration will serve.

as a self-regulatory device, it only works when the predators have alternative prey species they can switch to, and this they have among the many species of plants and animals living together in the African savannah. Likewise, in rainforests and on coral reefs, in mixed temperate woodland and on prairies/savannahs, most animals have alternative sources of food if one becomes scarce; so plagues are very rare in these species-diverse ecosystems, which are described as polycultures.

If we walked through an area of Mediterranean forest, in the first kilometre we would see examples of the vast majority of the species that lived in it. If a fire destroyed half of that forest, it is very unlikely that it would exterminate all the individuals of any species that lived in the whole of it. Certainly the fire would not cause the global extinction of a single species because no species is endemic (found only in)[78] to that half. If we walked one kilometre through a tropical rainforest of similar size, we would see only a fraction of all the species that exist in it. If people cleared and burned half of that rainforest, they could cause the global extinction of several of the species that lived in it, because those species were endemic to the burned half of the rainforest. That is why rainforests are so important to the biodiversity of the world.

The Arctic tundra is an oligoculture – that is to say, it has few species living together in it. Where predators are specialists at catching one prey species, they cannot easily switch to another when their principal food becomes rare. For example, snowy owls[300] feed mainly on plant-eating rodents called lemmings. Arctic foxes also turn to lemmings when they can. Lemmings are famous for their population 'eruptions', as the hordes that swarm over the tundra from time to time are called. Why they erupt is uncertain; it is likely to be a combination of population density triggering restlessness together with shortage of food. In other words, they have degraded their environmental. The

[78] *Endemic* is an abused word; correctly it means 'found only in', and should always be used in conjunction with the name or description of a geographical location.

lemmings' behaviour is exactly paralleled by human economic refugees.[79]

When conditions are favourable, lemmings breed faster than the owls can catch them, therefore lemmings become a plague. The only way snowy owls and Arctic foxes can control lemming populations is by these predators increasing their own population. But owls and foxes breed more slowly than lemmings; it takes several years for them to build up their population enough to reduce the lemmings. When the lemming population falls, foxes turn to other foods, but owls starve and their population crashes. A cycle of plague and starvation is the normal pattern in oligocultures like the Arctic. Examples of other natural oligocultures include deserts and pine forests both of which regularly suffer plagues: locusts and processionary caterpillars respectively. This is precisely the pattern of boom and bust that is normal in economic cycles where there is a narrow range of commodities being traded.

We human beings evolved in a species-rich world, and have become the most common and widely distributed large land vertebrate in Earth's history. We can feed on a vast range of plants and animals: rice and roots, whales and ants, leaves and fish – the list seems endless. We can defend ourselves against any predator, from driver ants to lions. More than that, we do not wait to be attacked first, we go out and destroy anything that seems to be a threat. Any living thing that is not directly useful to us is perceived by many as such a threat. We are reducing the world to an oligoculture comprising us and our food.

Large fields of single-species crops are obviously efficient farming, but they are at risk from plagues. In a naturally diverse ecosystem, bean plants (for example) are interspersed with bushes in which ladybird beetles lurk. A greenfly arriving on a bean plant stands a good chance of being eaten by a ladybird that is seldom far away. In a field of beans, a few greenfly will very quickly become a plague, there being no ladybirds nearby

---

[79] Polunin (1998: pp. 82, 176, 266 and 273). See also Ridley (1993: p. 105 *et seq.*) for a complication of this idea.

to eat them. The only way we can control such insect plagues is by pesticides. But pesticides do not distinguish between greenfly and ladybirds, and not only do greenfly breed faster but ladybirds suffer the fate of our poisoned sparrowhawks, so the cycle continues with oscillations of increasing amplitude.

Cities are oligocultures, so plagues, especially of diseases, are a real risk, and controlled only by modern medicines, which can be thought of as forms of pesticide. As the human population rises, and cities become ever larger, so the risk of plagues increases. More than that, the microbes that cause disease are becoming increasingly immune to antibiotics; Darwinian evolution is working on them as on everything else.

The one vital lesson to be learned from ecology is that diversity means stability. The four great categories of living things – green plants, plant-eating animals, animal-eating animals and decay or disease microbes – are fundamental to all ecosystems. The fewer species in each category, the more unstable is the ecosystem. A field of wheat, which may be called a 'corn desert', consists ideally (from the farmer's point of view) of one species of green plant and one plant-eating animal (Man); it does not even need decay microbes. Stability, even for as little as one growing season, can be maintained only by using fertiliser and pesticides. The tiniest change in conditions can precipitate a catastrophic change in the system – a single greenfly or rust spore arriving in a lull in the pesticide programme, a hailstorm, or a drop in temperature, and the crop fails. These same changes in a mixed woodland pass unnoticed because there are alternative species to take the place of the affected one. And we are cutting down the Earth's natural forests to make corn- or oil palm-deserts as fast as we are breeding.

We have seen that a pyramid of biomass represents the weight of living things in each of the three of the fundamental categories green plants, plant-eaters and animal-eaters. Their proportions are shown in Figure 10.4.

Now imagine a similar pyramid but, instead of biomass, we show the numbers of species in each category. We now have a pyramid of biodiversity. If there were 10 species of animal-eater, 30 species of plant-eater and 100 species of plant, the pyramid would resist the winds of environmental change (Figure 10.5).

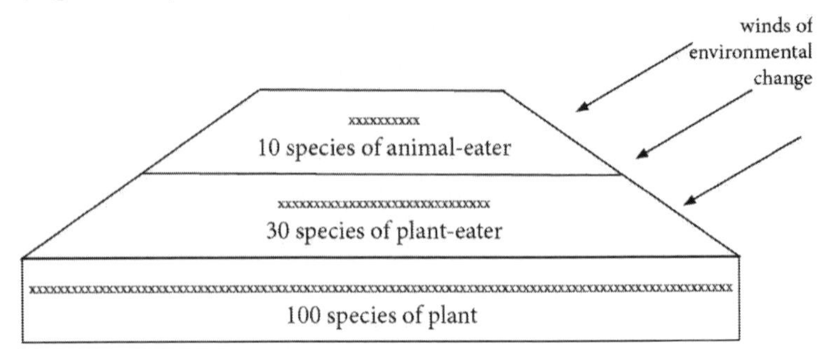


**Figure 10.5** A pyramid of biodiversity (polyculture).

A diverse ecosystem has great stability in the face of the winds of ecological change: several species can become extinct without it collapsing. How would an ecosystem with one species in each category fare (Figure 10.6)?

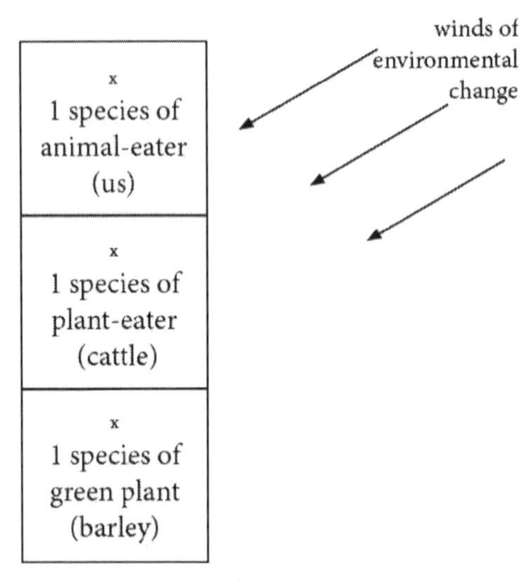


**Figure 10.6** A ‘pyramid’ of biodiversity (oligoculture)

One extinction would bring the ecosystem crashing down. Of course, these examples are simplified to absurdity but they illustrate the point that impoverished ecosystems are more vulnerable to environmental change than those that are species-rich. Consider two small towns, one with a mixed economy and the other specialising in raising and butchering meat. An epidemic of a meat-transmitted disease, such as bovine spongiform encephalitis, that persuaded people to become vegetarian would be disastrous for the second town, but the first town would survive economically, even though their butchers had a hard time.

**********

Whereas animal food consists of energy-rich organic compounds, plant 'food' consists of low-energy, simple inorganic compounds – they obtain their energy from sunlight separately from the material content. These inorganic compounds include carbon dioxide and water to make sugars, which are the starting point for most of their synthetic processes. Plants also require the following elements: nitrogen and sulphur to make proteins, phosphorus for DNA and to release energy, sodium and potassium to make cytoplasm, iron and magnesium to make chlorophyll, calcium to balance the acidity of tissues, and a handful of other elements in traces to make enzymes. Plants take in these elements as various compounds, some of which are abundant and others scarce. We should understand where these basic materials and energy come from.

There is no shortage of carbon dioxide in the atmosphere. Water supplies vary in volume and time, and different plants are adapted to its availability. Our food plants mostly evolved in temperate regions where seasonal rains were reliable; and we try to emulate this by draining swamps and irrigating arid regions.

There are 14,000 cubic kilometres ($km^3$) of annually available freshwater resources on Earth, and together we consume about 4,400 $km^3$ of them, of which 3,500 $km^3$ are for irrigation.[301] This works out as 500 cubic metres ($m^3$) for each one of us per year for irrigation and another 128 $m^3$ for all our other needs, such as drinking, cooking, washing and all the industrial processes that produce our food and toys. Each of us uses 90 $m^3$ annually if we have a daily bath. There is not space here to go into more detail about fresh water; it is enough to say that diminishing availability of water will become another cause of strife as our population rises.

The atmosphere contains 80% by volume of nitrogen, but plants cannot use it directly. However, certain bacteria can, and they convert the gas to soluble mineral salts, mostly nitrates, which plants can absorb. Nitrates are easy to make artificially, though the process needs much energy. We spread nitrates liberally onto farmland to promote the rapid growth of crops, but they are highly soluble in water and rapidly run off into rivers causing algae to bloom and then rot in a process called *eutrophication*. Sulphur is needed in smaller quantities, and it is amply available in the soil.

In ordinarily fertile soil, phosphorus is rarer than sulphur, but plants need more of it. They use it to release the energy that powers chemical reactions in tissues, and also to make nucleobases in their DNA and RNA. Vertebrate animals also need substantial quantities for their bones and teeth. Each ecosystem has one or a few main reservoirs of phosphorus and a useful question to ask is where it is. For example, in rainforests it is in the tree trunks, in the African savannah it is in the mammals' skeletons, in a coral reef it is in the fish, in temperate farmland it is adsorbed on to soil clay particles.[302] This is why taking a big tree from a rainforest or a big fish from a coral reef or game from a savannah lowers the productivity of each system. 'Sustainable harvest' from natural ecosystems is an urban myth.

Phosphorus[303] occurs in rocks, and it is released naturally by weathering. It is mined mainly in Morocco and there is only a finite amount of it in the Earth's crust. We are squandering the remaining stocks at a rate that has no thought for future generations. We put huge quantities onto our farmland and ensure that what does not run off into rivers ends up in our sewage works. There are other reserves of phosphorus in the Earth's crust but they are expensive to extract.

Sodium is abundant in the soil but potassium is scarcer. Sacks of farm or garden fertiliser often have three numbers printed on them, and these are the proportions of nitrogen (N), phosphorus (P) and potassium (K). A few other elements are needed in traces and are usually present in sufficient quantities.

**********

So much for plant nutrients. More importantly, we are losing the soil itself by damaging the protective plant cover. Take a spade and dig a hole on a sandy beach: easy. Now cut a hole in a lawn: more difficult to get through the turf. Now do the same in a woodland: it would take you a lot of effort to cut through the matted woody roots – it is hardly more difficult to dig a hole in a carpet. Roots bind soil particles together and prevent wind or moving water from carrying the soil away. A permanent covering of living plants over the land gives off water vapour and cools the air above it, allowing rain-bearing clouds to travel over it and drop their water far inland. Strip away the plants either by deforestation or winter arable crops, and the sun heats the bare rock dispersing rain-clouds and spreading the desert inland.[304]

Human history is an almost uninterrupted and dismal catalogue of ecological devastation, mostly caused by farming, followed by war and often cannibalism. Here are some examples.

**Australia**: a land already poor in plant nutrients but with a rich megafauna and plant cover when we arrived some 50,000 years ago. It did not take us long to cause the extinction of larger animals and burn off the vegetation. Now the continent is mostly desert.[80]

**Cañón del Chaco, New Mexico**: a rich and diverse culture dependent on agriculture which destroyed the productive ecosystem.[305]

[80] Flannery (1994) is devoted to explaining this.

**The North American Dust Bowl**: created from highly productive and species-rich prairies by simply ploughing sandy soil.

**Madagascar**: one of the swiftest and most profound biological catastrophes in the history of the world; most of it during the 20th Century – with 90% of its forests destroyed.[306]

**New Zealand** (and many other, now deserted, Pacific and Indian Ocean islands that once supported human populations): we drove to extinction most of these islands' species, which represented a fifth[307] of the species of birds on Earth, including the giant moa and the dodo. The appalling tale of destruction, cannibalism and extinction on **Easter Island**[81] is the classic model for the Earth's future, if we continue as we are.

**North Africa**: the breadbasket of ancient Rome, now virtually desert. Likewise the great **Sahara Desert** once teamed with game in semi-arid savannah; we burned it out before we learned to write, so scrawled what we saw on rocks.

**Near East**: the cradle of civilisation (farming, war and religion) that was once a fertile crescent; now it is a desert that, in the 21st Century, is still being fought over, though with renewed bitterness.

**Petra, Jordan**: capital city of the Nabataeans from 312 BC; flourished for a thousand years until it had destroyed its ecological base; now another desert.

**Persepolis, Iran**: formerly a mighty city, capital of ancient Persia, dependent on agriculture, now stranded in desert.[308] Percy Shelley's sonnet says all that needs to be said about a once fertile land.

---

[81] Lomborg (2001: p. 29 R 2, 3) makes the mistake of comparing closed ecosystems with open ones, especially in his dismissal of Easter Island: 'trade and transport act to reduce local risks'. Which island should the Easter Islanders have traded with, and which planet does he think the Earth is going to trade with? Easter Island as an ecological model of Earth is a good one. In fact, it seems that it was the exploitation of the island by Europeans that finally finished the destruction.

**Ozymandias**

*I met a traveller from an antique land*
*Who said: Two vast and trunkless legs of stone*
*Stand in the desert. Near them, on the sand,*
*Half sunk, a shattered visage lies, whose frown,*
*And wrinkled lip, and sneer of cold command,*
*Tell that its sculptor well those passions read*
*Which yet survive, stamped on these lifeless things,*
*The hand that mocked them and the heart that fed:*
*And on the pedestal these words appear:*
*"My name is Ozymandias, king of kings:*
*Look on my works, ye Mighty, and despair!"*
*Nothing beside remains. Round the decay*
*Of that colossal wreck, boundless and bare*
*The lone and level sands stretch far away.*

The list of ecosystems we have destroyed is long, and increasing yearly; however, there are some pockets of rainforest, seas and polar land that we have not yet turned into human waste. We, in the developed countries of the world, are actively supporting this destruction by buying rainforest products, such as palm oil derivatives and exotic hardwood; and by sponsoring scarcely democratic governments of those lands to do our will with arms and food aid. Insulated in our opulent cities from what is actually going on in the world, we are very near to destroying the global ecological base on which we depend.

It is not possible to preserve or conserve an ecosystem, for all are continuously changing. All we can hope to do is to protect the processes by which ecosystems came into being and by which they remain, for we depend on them utterly.

For them to function, all ecological systems depend on a flow-through of energy. In an exactly analogous way, all economic systems depend on a flow of money – much of it cycling, but also with import and export to other eco(nomic)systems. Since our destructive impact on the world is caused not just by our numbers but as much by our economic activities, we need to look at them, and to understand what would be the consequences of applying ecological principles to them.

# Chapter 11

## MONEY

By repetition, the phrase 'the London question' loses some of its punch. We may refresh its enormity if we try to put a price on carrying out such a vast engineering and social project. Of course, we would not try to build a replica of London in one place with all its ancient monuments, but would add the average kind of house to existing cities.

From a quick trawl online in September 2015, a 150 $m^2$, two-bedroom house of cheap design, but still conforming to UN requirements, built in Washington State, USA would cost USD150,000 plus land and furnishings. Allowing four people per house, out of the 8,000,000 (roughly the population of London or New York) requiring accommodation, we still need 2,000,000 new houses every five weeks. This would cost USD300 billion (300 thousand million), which is more than USD3 trillion (3 million million) every year. Remember that a million seconds is 11 days, but a trillion seconds is 32,000 years. That long ago, we could have met a Neanderthal Man.

Building USD3 trillion-worth of homes each year is the easy bit of the project, because the occupants still require, according to UN demands, an infrastructure of water and drains, electricity and heating, roads and rail; not to mention civic services, such as hospitals and fire brigades, schools and universities, police stations and town halls, post offices and telephone exchanges; all fully staffed with trained personnel, and also supplied anew every five weeks. I shall not begin to estimate what all these cost, but housing is clearly the cheaper part. Oh yes, and an average of one-in-four of these freshly housed and served people will need employment; that is 2 million new jobs every five weeks. Some of them will fill the posts listed above, when they have been trained, but more will be needed to create the wealth which the civic services spend. How can that wealth be created without using up the environment even more than it already is?

All this *before* we can start improving the lot of the 1.5 billion people already living in poverty. Which budget will these sums come from? Building the equivalent of a new London every five weeks costs more than mere money – it costs the environment. Anyway, what is money? In trying to answer this and understand each stage, I shall again take a naturalist's point of view, and start at the beginning. In this way, we will work our way towards understanding what money really is, rather than accept some shallow definition.

**********

In an early farming community, a man and a woman together could probably do everything that human beings could do: gather food, find shelter, evade predators and breed. More specifically, they could build and repair their house, knap stone tools, make bows, arrows and other weapons, treat injuries, find herbs, gather wild fruits, hunt wild animals, fish, herd, slaughter, butcher, till, sow, harvest, thresh and mill grains, dig and refine clay, throw and fire a pot, weave and sew to make and mend clothes, and smelt and cast metals.[309]

Clearly some people were better at particular tasks than others, so divisions of labour and barter soon developed. To know where they stood in the social organisation of their community, each person had to keep mental track, not only of the reputations and life stories of everyone they knew, but also who owed favours to whom, what those favours were, and how much were they worth in kind. Life soon became too complicated to remember all the details.[82]

In searching for suitable stones for tool making, our ancestors certainly came across lumps of mineral that did not flake or shatter when hit, but were malleable and became shiny. These are metals, and copper is the most common metal that

[82] Cartwright (2000: p. 303): wealth as an embodiment of indirect reciprocity; Buchan (1997: p. 22): money's origin in division of labour.

occurs naturally in elemental form as well as its ore. Copper appeared for the first time as manufactured objects in the Neolithic. Before long, people noticed that heating copper ore in a fierce charcoal fire turned it into the metal, a process called smelting. They soon discovered that copper could be hardened during smelting by the addition of arsenic or, later, tin. The addition of tin made bronze.[310]

Unless they found a large nugget or struck a rich seam of ore, it would take time for someone to collect enough copper to make an axe head of useful size. Almost certainly a man who already had an axe, or a woman who did not want one, came across copper ore and smelted it into portable pieces of the metal. It needed little imagination to exchange these surplus pieces for other goods that they wanted and which were available. Once copper, or any other substance, had made the transition from being the raw material for making something useful to a token that represented other goods and services, it had no real value until it was exchanged for them.[311] This system of exchange worked well enough while the population and primordial economy grew, so long as new veins of copper were being discovered.

There was little obvious merit in hoarding copper for later exchange in general trade when other people needed it to make something useful like an axe head; obvious, that is, to those of cooperative or generous nature. In contrast, less socially responsible people could deliberately push up the value of copper, in terms of what it could be exchanged for, by hoarding it. Here we are getting an early glimpse of the distinction between need and greed, and these two motives now had a currency: money. There is nothing modern about cornering a market in a commodity – squirrels and chimps do it. Money takes on a life of its own when, having been separated from direct use, it begets the desire to possess more of itself; and this property is inherent in it simply by its being money.

Gold and silver also occur naturally as elements, but are too soft to be of much practical use. Like copper, they can be melted or hammered into shape and polished; they retain their shiny surface rather better than copper, and easily take a stamp. They also survive repeated melting without oxidation, and can be buried without too much tarnish.[312] These properties, together with being both rare and comparatively useless, made gold and silver suitable as tokens to represent larger quantities of useful copper, and also other goods or services. In this way, gold and silver became true money, which is anything that is worthless until it is exchanged for something a person wants more than what they have (even if that 'thing' be social position). The invention of money made it no longer necessary for people to remember what they owed and were owed for goods and services rendered; they could assess and cancel the debt at once.

But there is a deep flaw in this concept of money. A thatcher helps his neighbours to roof their cottage, and in exchange they give him two legs of lamb. Living under a newly thatched roof and eating lamb are the two ends of a process; in effect, barter annuls a contract by satisfying two needs. But payment in money satisfies only one end because money can continue its life and be exchanged for something else.[313] Thus, possession of money never satisfies; indeed, it creates desires that no amount of it will fulfil. A failing that was identified in the phrase 'the love of money is the root of all evil'.[314]

If the neighbours had had a bad farming year and could not repay the thatcher at once, he trusted them to do so the following year. The price of thatching a cottage in the autumn in return for payment in lamb the following spring would have been higher than if the debt had been settled at once because it included the risk that the neighbours might default – perhaps through illness or other unforeseen circumstance. To cover this risk, the thatcher insured against it by adding an extra leg of lamb to his bill. It was not that the thatcher did not trust his neighbours; by giving them credit, he had no choice but to trust them. His reasons were purely practical: he had to cover the risk of the few defaults he was likely to encounter in his working life. This aside, money normally removes the risk in waiting for repayment in kind, and thereby also removes the trust. Trust demands commitment and involvement, both of which require social effort. But neither is needed when debts are discharged with money.

For example, in the southern Spanish village where I once lived, tradesmen were almost irritatingly reluctant to send out bills – irritating only to those imbued with the monetary values of northern Europe. A tradesman hastening to demand payment was seen as belittling my credit worthiness; conversely, my hastening to pay, hinted that I had a suspicion that the tradesman was poor enough to need the money urgently. Such interactions were largely a matter of 'face', which life in a Spanish village protected by social effort. This is true all over the world except when the love of money is in control. The seemingly endless hours of village gossip lubricates the microscopic economic wheels of village life. But, like most social interactions, there are undercurrents. I feel slightly inferior to someone to whom I owe money; this alters my body language to reflect a tiny posture of subservience, and this gives me a minute disadvantage when the two of us, side by side, chat up a girl, because she is expert at unconsciously picking up such unconscious signals.

Money was particularly useful when dealing with strangers because people could now carry with them a statement of their worth. It was clearly more sensible for a trader to do business with someone whose reputation he did not know but had money with them, than a stranger who did not. It is no coincidence that the first known coins (dating from the 7$^{th}$ Century BCE) were found in Lydia, which lies at the meeting point of Asian and European peoples.[315] Midas and Croesus, whose names have gone into many languages, were Lydians of that time.

**********

In enabling instant payment, money creates a Darwinian market: tradesmen who lowered their prices for upfront money would have been at a commercial (= selective) advantage over those who did not. On the other hand, there are occasions in which it would be convenient to defer payment, even if it costs a bit more.

A woman corn merchant can initiate the process of giving credit by perceiving that a farmer is hungry. She offers to buy his harvest as soon as she sees it growing.[316] There is no need for her to insist that he should cultivate it properly because she knows that his reputation is important to him, and that he can scarcely run off with the farm. The price she offers is lower than he would get later in the summer because she has taken on the risk of the crop failing, or there being a glut which would lower prices. There is also a chance that such fluctuations would favour her, but she does not mention that. Trading in futures is surely as old as gambling. Is there a difference? From this trite beginning, the vast edifice of financial services bloomed, offering ever more sophisticated instruments. These instruments performed a service only so long as they facilitated deals; as soon as their purpose became a means of making money, they degenerated into parasites, and ceased to contribute to common wealth.

Usury, in the sense of lending money at interest, is recorded from 4th Century BCE Greece where interest rates were set at 10% for ordinary business and up to 30% for shipping and allied risks. The word 'usury' is sometimes used pejoratively when someone wants to signal that they disapprove of the rate of interest. This is not justified when the contract to lend/borrow is clearly understood by both parties and freely entered into. If there is coercion, then the transaction becomes extortion, not usury.

A balance between how urgently the borrower needs the money and how urgently the lender needs the interest on lending it, sets the interest rate. Money does not have feelings, and moneylenders are not obliged to be charities. Even if they appear to be generous in charging a low rate, they may well have in the back of their minds that they are not only taking on a risk but also building up a reputation to attract the customer back again on a future occasion. Another motive of moneylenders is to encourage the borrower to work hard, build up a business and trade with the lender. If the borrower did not do this but bought luxuries instead, the lender would have made an error of judgement and might as well have backed a slow racehorse. In neither case has the lender/gambler a good reason for demanding repayment. The present (September 2015) monetary crisis has mutated into precisely this form. Greece has borrowed €330 billion (= 180% of its gross domestic product (GDP))[317] and not met its repayment dates. It is now asking for more loans to save it from bankruptcy.

Usury was controlled from the earliest times. The Old Testament[318] prohibits it when lending to the poor or within families but allows it with foreigners, which may be one cause of gentiles' suspicion of Jews. For Christians, this rule became prohibition for priests in the 4th Century CE and more generally a millennium later;[319] meanwhile Mohammed copied the Judeo-Christian tradition for Muslims. If we cannot lend money, we may as well hoard it, which restricts commerce, so usury soon returned in business communities.

**********

Borrowing and lending money becomes more problematical when it is between generations. In an impoverished society in which two-thirds of a man's children die before he does, he needs 12: on average 6 boys and 6 girls. The girls are unlikely to have control over any money they might accumulate, so he sells them (bride price). Two-thirds (four) of his sons die and that leaves him with the two that he needs to look after him in old age. In modern industrial societies, we share responsibility for looking after pensioners by paying taxes and transferring the practicalities of this caring to the state. But such a system works only when economies are expanding and there are more and more children growing up to be the taxpayers of the future. The system breaks down when the workforce diminishes and the number of pensioners rises: hence the present-day (2015) worries about ageing populations throughout the world.

In 1881, Joseph Logan put his belief in the ever-expanding population and resources into distressingly direct effect. He was a shipbuilder of Scottish descent who had emigrated from northern England to Canada with his family, which consisted of his wife and the surviving 10 sons of their 14 children. Logan sent for his eleventh son on the boy's 21st birthday, as he presumably had the previous 10, and presented him with a bill for 537.50 Canadian dollars (CAD) (a huge sum in those days), being the carefully recorded expenditure of the boy's life – even the doctor's fee for bringing him into the world – all unquestionably correct. In a flowery speech praising his own efforts, and scarcely acknowledging his wife's, who had brought a CAD20,000 dowry to the marriage, Logan required his son to repay the debt at the earliest opportunity, charging annual interest of 6% on the balance. Small wonder that, having paid this debt and that of his brothers, the great Scottish-Canadian (and naturalised US citizen) naturalist and illustrator Ernest Thompson Seton changed his surname.[320]

The present human population tsunami would seem to solve the pensions problem; there will be plenty of young workers to look after the retired. But it is not just a question of numbers. Because of our excessive demands on it, the environment will stop providing us with food, water and waste disposal, unless we all accept a vastly reduced standard of living, and reduce our numbers.

**********

Thieves and cheats have always been with us, and the invention of money supplied them with new opportunities. Two of the easiest ways of cheating in a deal that involved metallic coins were by short weight and adulteration. Few people carried scales to weigh each nugget of metal they dealt with, nor could they easily determine whether it was pure. A standardised piece of metal with the chief's mark on it would be perceived as being more reliably pure and of a stated weight than one without.[321] Marks are more easily stamped onto a nugget if it were flattened like a coin. But cheats could still clip bits of metal off the edges – a practice that continued well into the 17th Century. This fraud was so pernicious that mints began to mill coin edges so that clipping could be detected. Cappadocian rulers were guaranteeing the quality of their silver ingots in the third millennium BCE.[322]

Weight for weight, incompletely smelted copper ore is less useful or valuable than pure copper, so a cheat could enclose ore, or some other cheaper material, in a piece of pure copper to adulterate it. The Bronze Age intellectual Archimedes lived in 3rd Century BCE Sicily. King Hiero II of Syracuse asked him to find out if a crown, supposedly made of pure gold, had been adulterated by the goldsmith. Archimedes noticed, as he lay in his bath, that his legs required less effort to raise than they would if unsupported by water. From this observation he devised in his mind a simple test for the purity of gold, or any other homogeneous material of whatever shape. At the moment of inspiration, Archimedes leapt from his bath and ran into the street shouting *Eureka! Eureka!* ('I have it! I have it!' – the solution to the problem), but neglected to dress first, and was presumably misunderstood by passing citizens. Back in his laboratory, Archimedes simply weighed the crown in air and again when immersed in water and suspended from the scales by a thread. The second weight was always less than the first because of the upthrust of the water on the immersed object. This upthrust is the weight of water that the object – in this case a crown – had displaced, which is proportional to its volume. He repeated the measurements with a sample of pure gold and found the ratio to be different from that of King Hiero II's crown. Therefore the crown *was* adulterated, and presumably the goldsmith's retirement was immediate and brief.

It was a pity that a similarly simple test was not applied to the mortgages that were adulterated and passed off as pure AAA in the 2007 USA financial scam. With today's human population some 20 times what it was in Archimedes's day, the value of an individual person's life is one-twentieth of what it was then (inflation?). Why should not those who perpetrated the sub-prime fraud experience pension rights similar to those of King Hiero's goldsmith? The answer, of course, is that many 21st Century leaders in business and politics approve of the fraud, whatever they said in public, because it raises the GDP. So also did the outrages of 11 September 2001 and the rebuilding of Ground Zero. There is madness in this view of money.

From earliest times, chiefs or governments have adulterated and debased the money they stamped,[323] most often by amalgamating the precious metal with baser sorts, such as nickel or copper, arguing that the process made the coins more durable. Such alloys turn the real value of a coin as a quantity of metal, into a token for the exchange of goods and services. Once this boundary has been crossed, it does not matter what metal or impurities the coins contain, nor what form the currency takes, because the coins are only tokens. Cowries and other shells suitably drilled (wampum), stamped leather, fixed stones and even paper were all tried in ancient times.

In the 3rd Century CE, Rome reduced the silver content of its coins to as little as 4%. Because the Roman government no longer needed to buy rare silver to make coins, they produced many more of them, but doing so did not create more eggs, lamb, thatching straw, and so on, nor did it create more labour to produce and handle these goods. Therefore more money represented the same amount of goods and services as before, and this is inflation. The English penny contained the same weight of copper for 200 years until 1344 CE when it was slightly reduced, and again 7 years later.[324] However, this did not matter because the mint did not produce more pennies, so people continued to have faith that the penny coin represented the same value in goods and services as previously. Nor does the conversion of coins to paper cause inflation, so long as the nominal value against available commodities and labour remains unchanged. Unhappily, because paper notes can be written out or, later, printed much more easily than metal coins can be minted, governments are tempted to produce more than the economy could support. The first recorded issue of paper notes was in China in about 800 CE, and the reason was a shortage of copper. ('Cash' was the name of their base metal coin, and it survived until 1912, after two millennia of circulation.) Within 200 years of introducing paper money, Chinese inflation was galloping, and hyper- a century later. They eventually abandoned it after 500 years of bitter experience. Today, money is reduced to pixels on computer screens. Under the euphemism of 'quantitative easing', the US government issued USD84 billion *per month* of electronic figures on screens during the financial crisis of 2008–2013. This was not inflation, they said, because the economy grew by at least that amount. Meanwhile, the natural environment, on which we all depend, shrank; however, they did not say so because it would have been bad for business.

Paper money in 17$^{th}$ Century Britain had a different origin: goldsmiths let people deposit valuables in their safes and issued receipts for reclaiming them. A depositor could then show these receipts to someone else as evidence of his ability to pay – his credit rating was higher than if he had no such paper. The receipts quickly evolved into banknotes and cheques, which had to be backed by coin or other treasure in the goldsmith's vaults. Issuing notes converted goldsmiths into banks, though some banking activities, such as exchanging and storing coins, were already ancient – there is evidence for them going back to the invention of money. The Bank of England was founded in 1694 and a decade later parliament legalised the negotiability of goldsmiths' notes; that is, the bearer, rather than a named person, could reclaim the value stated on it. This is why English banknotes today (2015) bear the signature of the chief cashier under the words, 'I promise to pay the bearer on demand the sum of …' I remember paying my college fees in 1956 with a white £5 note printed in swirling black characters, and having to add my signature to several already on it. The college secretary held it suspiciously to the light to see the watermark. That white note was the first either of us had seen; and it was almost the last.

Apart from money or bullion deposited in the banks' vaults, the safest asset against which a bank could make a loan was, of course, land because the borrower could not abscond with it. Title deeds were essential prerequisites for mortgages. Banks issued notes not only as receipts for property that depositors had pledged personal loans against, but also as loans against the bank's own coin or property. Issuing notes against the expectation of repayment is entirely different because it is borrowing against the future rather than the present.

**********

As the human population increased, there were more man-hours for labour, and more materials were extracted from the environment: food and water, wood for fuel and mineral ores and later coal then oil. To allow the efficient exchange of these commodities, money circulating in the economy had to increase in proportion to economic growth, if only to prevent cumbersome and inefficient trade by barter, or deflation in which money becomes more valued than commodities thus causing prices to fall. When this happened, people hesitated to spend money in case prices fall further; this depresses economies.

When paper money has to be backed by gold, trade is restricted by the amount of the metal held in banks. The 16th Century robbery of New World gold by the Spanish conquistadores, and discovery of a mountain of silver ore at Potosí in what is now Bolivia, stimulated both trade and inflation. The 19th Century gold rushes in California, Australia and Alaska supplied sufficient amounts of the metal to back American and European economies, which then grew as a result of this injection of liquidity. When these sources declined, economies slowed, so thought was given to alternative ways of backing paper money. Just as hoarding copper prevented other people from making axe heads, so hoarding gold slowed world trade. USA and France were specialists at this and so helped create the economic recessions of the 1920s and 1930s.

Approximately 163,000 tonnes of gold have ever been mined: 86,000 tonnes are in private hands, mostly as jewellery; 30,560 tonnes are held in national or central banks; 26,400 tonnes are invested as bars and coin; and 20,000 tonnes are in industrial use. Gold mining companies have declared that there are about 50,000 tonnes still in the Earth's crust. It is an interesting idea that the total volume of gold on Earth would fit into one large ship. During World War I, gold coins were steadily withdrawn from circulation, and new notes issued, so quietly bringing an end to gold as backing for currency – the gold standard.

**********

All material goods and the energy needed to handle them are ultimately extracted from the environment. I experienced a startling illustration of this by travelling towards a Moroccan town through a sparse forest. Soon the trees became smaller and then disappeared; next there was a belt of scrubby juniper and that gave way to ankle-high shrubs, pasture and then to arable fields. Immediately around the town was bare earth and rubbish. The outer zones of the town were crowded with poor housing served by unmetalled roads, open drains and the uniform arcades of simple workshops and stores; nearer the town centre were villas with gardens and sealed roads. The town centre itself was obviously wealthy with European shops, street lights and much traffic. In the very centre of the town was the local lord's kasbah (citadel). I entered it through a gatehouse, crossed a courtyard and passed through rooms of increasing luxury until, in the very heart of the citadel was a locked door: the treasury. Behind it lay the 'standard' with which the lord backed his reputation for wealth.[325] For all I knew, it may have been empty; but that is not the point I want to make here. My point is that the countryside for many kilometres around the town had been stripped of its materials: animals, timber for construction and fuel, grass and minerals from the soil in crops, plus metals mined from more distant sources. It was the poor people living on the edge of town who had done the stripping and manufactured the goods, which the lord had then 'taxed' to furnish his kasbah and fill his treasury.

The similarities between this Moroccan (and many another) town and an ants' nest are too great to need comment. The tentacles of our modern cities reach out across the Earth, and this pillage is energised by oil, though largely by charcoal and donkey power in the 1990s Moroccan town. As well as bringing in money, trade satisfies some deep-seated desire in human beings.[83]

Apart from fulfilling the need to trade, the desire to accumulate money frequently becomes obsessive. Those manufacturers, service providers and traders, in whom such a desire has taken root, continually seek to stimulate consumers' appetite for more and different products.[326] This is fair enough, but so strong is the lure of money that people are prepared to take shortcuts. For example, from early times competitors were

[83] O'Hear (1999: p. 160): trade gives people a high idea of their personal importance. Eibl-Eibesfeldt (1996: p. 189): and induces bonding.

seen as a threat, and traders expended much effort in cornering a market and stifling competition.

To combat a particularly virulent episode of $18^{th}$ Century mercantilist protectionism, as it was called,[327] a Scottish academic called Adam Smith (1723–1790) described a theoretical model of how money flowed around a trading community.[328] He called his model *economics*. He showed that the wealth of a nation was not necessarily measured in the amount of money or treasure it had, but in its capacity to increase that wealth by trade. His model was based on the idea that an individual who intends only his own gain, competes with other such individuals for customers' money. Those who are more efficient at supplying customers' wants succeed, and those who are less efficient go out of business – Darwin's idea again. The overall result is that the community as a whole obtains goods and services in the most efficient way. This was an enormous relief to those who felt guilty about the conflict between working for themselves and working for other people. Now, at a stroke, Smith had explained how selfishness could be beneficial to other people.

It is widely thought that Smith coined the idea that a business person was 'led by an invisible hand'[329] to promote public interest. He did not. He probably lifted the phrase from Daniel Defoe's *Moll Flanders*, and used it three times in all his surviving work, not one of them in connection with free market capitalism.[330] In fact, Smith was describing the operation in business of what we now call Darwinian evolution. Like the phrase 'survival of the fittest', 'invisible hand' is meaningless, though it has caught ill-informed imaginations. In James Buchan's brilliant words, 'For the moment, the invisible hand is a projection of the yearning for coherence on to some supreme agent, on a father in a fatherless world.'[331] The reality, which we are so reluctant to face up to, is that the ability to reason in a reasonable but unreasoning world makes fathers of those who do.

**********

There are three fatal flaws in free market capitalism:[332] (1) the idea that the resources of the world are infinite and growth can continue indefinitely; (2) the sort of mentality that believes that the world will be richer when the last elephant dies because the price of ivory goes up;[333] and (3) the tragedy of the commons.

Commons.[84] When a landowner manages his own animals, he eventually notices that his land can support, say, 100 adult sheep plus their lambs in spring when grass is growing quickly, and that adding one more ewe *decreases* the overall production of wool, milk and meat by, say, 2%. Being sensible, the landowner restricts his flock to 100. If, on the other hand, the 100 sheep belong to 5 families each owning 20, it is worthwhile for any one family to add one more ewe to their holding, because each added animal represents 5% of their flock less one-fifth of the common 2% loss in production per ewe added. In practice, if more than one ewe is added to the critical number, the loss rises steeply because overgrazing leads to erosion. On our assumed figures, if a family adds one more sheep, it would gain by 5% less one-fifth of 2% (= 4.6%), though the village as a whole would become poorer. The fewer sheep each family owns, the more worthwhile it is for them to add another because the greater would be their percentage gain. In other words, the poorer a people are the poorer they will get under this form of land management. This is a positive feedback cycle exacerbated by aid from richer countries. Enclosure of English commons in the 18th Century had more to do with profit than class issues. Nowhere is the tragedy of the commons more evident than in the oceans of the world, not just for fishing but especially for dumping horrendously toxic chemical and radioactive waste.

In 2004, the charity Christian Aid launched an appeal called *How to recycle a goat*. The instructions read:

---

[84] Hardin (1968: p. 1244). Grove & Rackham (2001: p. 88) and Monbiot (1994: p. 140) argue that commoners learn to manage their land efficiently – this is true in a few specific cases. However, the vast majority of common lands are destroyed when regulation is lifted, as it inevitably will be when population pressures rise; see also Anderson (1971: p. 22), Cohen (1995: p. 257 *et seq.*) and Ridley (1996: pp. 87, 103).

Start here:
Give a widowed mother a goat.
The goat produces milk. Her children don't go hungry.
The goat produces manure. The widow sells more crops.
The goat produces more goats. The widow keeps a goat.
The widow gives a goat back.
Back to the start.

This mantra reveals a profound ignorance of reality: there is no mention of what the goat would eat – its food would almost certainly come from common land. Nor is there any mention of *more* children and *more* goats but the same amount of land and less water. This is precisely why aid has so drastically damaged much of the fragile arid and semi-arid regions of the world, so that they are now producing food at a rate far below their potential. I argue in the next chapter that campaigns like *How to recycle a goat* are little more than social displays of 'look-at-me-caring'.

**********

Economists have a problem. The subject matter of mathematics, physics, chemistry and biology remains unaltered, however we study it. But the ways in which we study these subjects has changed out of all recognition since the Renaissance. On the other hand, economies are a human activity and so are changing all the time, but the ways in which they are studied have changed little since Adam Smith and other founding fathers laid down the basic principles. Unfortunately, being based on fundamental misconceptions, these principles are deeply flawed. Indeed, human understanding of economies has barely begun.[85] Until economists start to consider economic contraction as a desirable aim, their subject will remain of dismal repute.

Money fails because it is required to perform two mutually antagonistic functions: it is used as a medium for the exchange of goods and services, and also as a store of wealth. A medium

---

[85] Buarque (1993: pp. 73, 125–129 and 151); Buchan (1997: p. 182) for an excoriating view of economics as an academic discipline.

for exchange must keep circulating; a store of wealth is, by definition, static.

The first backing for money, metal, failed because the amount of it in circulation was limited and it could be hoarded. These facts restricted economic activity too severely to keep pace with expanding populations. The second backing for money, confidence in continued growth, allowed increasing economic activity though it was vulnerable to inflation, bubbles and crashes. We are now well within sight of an end to growth, though politicians and business people generally seem incapable of understanding this; or, if they do, they will not say so.

We spend money on material goods and services by other people. All materials come from the crust of the Earth and extracting them involves disrupting the environment somewhere. Every service rendered involves the expenditure of energy, all but a little of which is also collected from the environment. Therefore every unit of currency we spend degrades the planet Earth. Therefore modern money needs to change its backing while there is still time to do so in relative calm – calm relative to the ecological cataclysm that will follow inevitably from not doing so.

In view of this argument, which is based on thought rather than feelings or belief, it is hard to understand why influential people in politics, business or the media insist on continued economic growth. And the really frightening thing is that most of us are encouraging them to do so. It is as if we are possessed by devils and running violently down that hill near Gadara.[334] Since the vast majority of us are neither stupid nor evil, why are we possessed by admiration for expanding economies? The answer is clear: we behave as if we were merely marionettes dancing on the strings of our DNA; and that command is *more*. We will have to change our economic behaviour one day, and the sooner we start thinking, as opposed to feeling, about it, the less cataclysmic those changes will be.

Though I shall make the bulk of my proposals for actions we can take to manage the human population tsunami in later chapters, it is convenient to describe here those that deal with money.

We cannot do without money, and it cannot exist without being tied to a standard. That standard should be a commodity that limits economic activities to truly sustainable levels.[335] Having established that neither metal nor confidence in growth are suitable standards with which to back a sustainable monetary system, I propose that a new currency be established, and that it be backed by reduced carbon (reduced in its chemical sense). From its relative atomic mass, 12 grams of pure carbon would seem to be a convenient unit (or a kilogram – it really does not matter), and it might be called the *Carb*. We will not be in a position to understand this proposal unless we are familiar with the simple chemistry described in Chapter 3.

Carbon is not only a fundamental component of all organic materials and hence life, but is also an excellent energy vehicle. The chemical energy in the bonds between its atoms can be released by simply burning it. That energy can be replaced by photosynthesis in sunlight: the process is exactly like recharging a battery.

Carbon and its compounds can be thought of in five forms: (1) elemental, that is, pure as in diamonds and graphite, and nearly so as coal; (2) reduced – also called hydrocarbons – as in oil and natural gas; (3) partly oxidised as in, for example, alcohols; (4) carbohydrates in which the proportion of hydrogen to oxygen in the carbon compound is the same as it is in water ($H_2O$); and (5) fully oxidised as in carbon dioxide ($CO_2$) and limestone ($Ca(CO_3)_2$).[86] When reduced, carbon compounds are fully charged with energy; when oxidised they are 'flat'.

All countries, or regions that share a currency, have carbon, even if their only source of it is in the air as carbon dioxide. To increase their wealth, all they have to do is to reduce carbon to usable form by allowing plants to grow. Some countries already have huge reserves of reduced carbon as coal, oil, gas and

[86] A purist might object that bicarbonates also contain hydrogen.

methane hydrate; and others as standing timber or fish stocks. Each country, or group of countries that share a currency, can evaluate its capital wealth in reduced carbon, and allow only that amount of currency to circulate. Burning fossil fuels, cutting down or squandering other carbon-rich components of the world's patrimony for short-term political or economic gain would diminish the currency a country could circulate.

Hydrogen burns giving off heat as H–H is converted to H–O–H (water). Thus hydrocarbons, for example, petrol (mostly nonane $C_9H_{20}$), contain no oxygen so both the carbon and the hydrogen can be oxidised to release energy. Ethyl alcohol is a partly oxidised carbon compound with the formula $C_2H_5OH$. When it burns, it releases energy from both its carbon atoms but only from four of its six hydrogen atoms because two of them will remain combined with the oxygen atom already present in the molecule. This is why the brandy on your Christmas pudding burns with a cool flame.

Most organic compounds grown on farms and in forests are carbohydrates with the general chemical formula $C_nH_{2n}O_n$. Clearly there is no energy to be obtained from the hydrogen when the material burns because it is already fully oxidised. But plants often convert the carbohydrates they manufacture to oils, and they do this by reducing the oxygen in the molecules. Fats and oils are insoluble in water and so easier than sugars for plants to store.

**********

The idea of carbon underpinning a currency is not new. Oil is the effective backing for the currencies of several oil-rich states. Economies throughout the world depend on the price of oil. The various factors that influence this price include: tax, demand for actual use rather than hoarding, production costs, fraud, war (usually seeking to control the resource) and speculation. The last is the most continuously destructive of market stability, and also the least predictable.

Some governments are edging towards environmental concern by introducing a system of 'carbon credits' in which businesses can buy a certificate from their governments that entitles them to release one metric tonne of carbon, or carbon dioxide equivalent, into the atmosphere. The number of certificates is limited by the 1997 Kyoto Protocol which 37 countries signed.[336] The targets were based on 1990 levels of carbon emissions, and varied between regions. The USA declined to take part, except to plead for more emission. (At least they were consistent with their stance at the UN Conference on Environment and Development held in Rio de Janeiro in 1992. Unlike most participants in either conference, the USA was also honest, as a cursory inspection of what has happened since either conference shows.[87])

The limitation of carbon credits created a market, and certificates could be traded: those businesses that used less than their allocation could sell them to others that used more. These measures were aimed at reducing carbon dioxide emissions. Jim Hansen, the world's pre-eminent climate scientist, has carefully described the full horror of how these measures are being studiously ignored by governments and commerce. He claims that the simple and certain consequences of continuing to burn fossil fuels at present rates are that sea levels will rise by at least 6 metres *when* the West Antarctic ice sheet melts,[337] which is likely to be within 50 years. That event will precipitate general ice cap melting and a sea level rise of 75 metres.[338] I know of no testable evidence that can be used to contradict his prediction.

No single-solution measures will diminish burning of fossil carbon sufficiently to reduce man-made $CO_2$ emissions. The only effective way to avoid the environmental catastrophe that is facing us is to link carbon to the global monetary systems. Existing measures would be entirely replaced by changing to a Carb currency. Three major effects of it would be that: (1) all citizens would be made constantly aware that every economic activity is environmentally destructive; (2) wealth, in the form of carbon, could be grown or simply destroyed by burning to

---

[87] To work through each of the 27 Principles of Rio 1992 is depressing in that their wording is exposed in this book as largely meaningless; an analysis would occupy too much space here. Kyoto 1997 set targets that can be inspected at http://en.wikipedia.org/wiki/Kyoto_Protocol.

release energy capable of doing work; (3) there would be a rapid slowdown in global economic activity with concomitant environmental recovery; eventually, economic activity would reach a level that the biosphere could support.

**********

In the past, people living in countries whose currencies were losing their value often insisted on payment in a stronger foreign currency – usually the US Dollar – for particular transactions, and kept their own local currency for domestic expenditure. They also sought to transfer their savings into that stronger foreign currency as quickly as they could. In effect, they were using the local and inflating currency as a medium for exchange, and the more stable foreign currency as a store of their wealth. In the same way, the Carb should be officially introduced to circulate beside local existing currencies. As all existing currencies are based on the expectation of future growth and inflation, they will inevitably weaken and be replaced as economies slow down to sustainable levels. In this way, the Carb would become a medium for exchange and its backing – existing stocks of reduced carbon – a store of wealth.

Inflation of a currency is caused by governments issuing more of it than the economy can support. It is a device whereby a government takes money from those who save and gives it to those who borrow.[88] With the Carb as a currency, economic activity would be controlled by the balance between inflation and deflation, which, in turn, depends on the balance between demand and availability of carbon. A carbon-standard currency and global economy based on it has the advantage that, by its

---

[88] Consider this: I lend my neighbour £100 to invest in a commodity. I charge him 4% interest for the year but, during that time, inflation has risen by 5%. My neighbour sells his investment for £105 and pays me back £104. What I could have bought with £100 a year ago, I now have to pay £105 for but I have only £104. I have lost 1% and my neighbour has gained 1%. If there had been deflation during the year, commodities would have been less in demand than money, so their prices would have fallen. My neighbour would have lost money on his investment and I would have gained because my original £100 could now buy me more than it would have a year ago.

nature, it is self-regulating. Thus there will be little need for stringent controls and interference with civil liberties, much as some creeds may desire them.

Introducing the Carb would be complicated but not difficult. A small 'Club' of wealthier nations could adopt the Carb as a currency for trade between themselves when dealing in certain defined commodities. Each nation would continue to trade other commodities and with other countries in existing currencies. Once the 'Carb Club' is established, and the reasons for its existence are clearly perceived, its inherent stability would attract more nations to join it. Other inducements would include the fear of economic isolation from the main producer nations. Governments that cling to unsustainable growth-based currencies would be demonstrating their contempt for the well-being of future generations, and be judged by their electorates accordingly. Such a stable currency as the Carb will also be anathema to speculators. By steadily increasing the goods and services that were traded in the Carb, it would eventually become the currency of global trade, other currencies eventually inflating themselves into oblivion.

**********

In addition to this change, three other monetary regulations should be put in place. First, amassing wealth has two immediate effects: it divides society by giving cause for feelings of inferiority and envy as well as superiority and smugness. There is strong evidence that nations within which there is great disparity of wealth between the richest and the average score badly in a number of positive social indicator fields, such as feelings of trust, life expectancy, numeracy and literacy.[339] On the other hand, such nations score highly in negative social indicators such as mental illness, drug use, homicide rate, infant deaths, obesity, high school drop-outs, teenage births, abortions, children's experience of conflict and the ratio of prisoners to population. And this is only in developed countries.

A way to reduce wealth disparity is not by taxation but by deciding on a maximum wealth that any one person can control. For all practical purposes, I cannot tell the difference between

£5 million and £10 million – not that I am ever likely to be in a position where I would need to. This suggests another way of slowing down economic activity, and it is to prohibit any one person from controlling more than, say, £10 million or its equivalent in other currencies. The surplus should not be taken away in taxes, but the law should simply require that the owner dispose of the surplus during the year after which they acquired it. This is what some ultra-rich people do anyway: they set up charitable foundations in which they play a part during their retirement. The Bill and Melinda Gates Foundation is an example. (Misguided, in my opinion; see later under measures to reduce our population.) Other people might simply wish to spend their surplus money on ephemeral consumables; they would then be judged by their actions in squandering the Earth's resources. Helping poorer people, funding research, supporting schools and colleges are all likely to become a common form of display: most normal people feel good when they are seen to work hard and then donate lavishly. The worst thing would be for any form of government, having set the wealth ceiling, to interfere in this redistribution. Evading this law by attempting to hide money in other regimes would necessarily attract penalties – both for the individual that did so and also for any institution or county that aided them.

While dealing with money, another reform that is urgently needed is to do with reducing both unproductive administrative work and crime. In this age of ultra-fast electronic communication and accounting, there is no good reason for paper money, or even coins. Indeed rechargeable plastic notes with personal identification on them are already (2013) available in Europe, and touch smartcards for small sums elsewhere. Those responsible for introducing the €500 banknote in 1992 may well have been motivated by the pure and honest attitudes of the Swiss who often do business with CHF400 notes; but I have less worthy suspicions. Anyway, the €500 note soon fell into disuse, to be followed closely by the €200, and even the €100 note had become rare by 2010. The days of criminals' suitcases of grubby bank notes are coming to an end: it is not easy to move sums of money in excess of USD5,000 other than electronically – even cheques are being phased out, but probably on economic rather than security grounds. Why should we not go the whole way and abolish all paper and coin money, and make every transaction electronic and therefore traceable? Software that would facilitate transfer of any sum, small or large, can easily be installed in cell phones. Records of electronic transfers would enable errors or fraud to be detected, corrected and punished more easily than in cash transactions.

With a major decrease in the gap between the rich and the poor, differential scales of taxation would become unnecessary. A flat tax rate of 10% could be paid directly into the treasury on *every* transaction: especially movements from one bank account to another, stocks and shares, houses and businesses, and particularly loans and interest. This would slow down economic activity and give real meaning back to money as a medium for exchange rather than as gambling chips and competitive display.[89]

When I kept a shop, I noticed that need was usually the last reason customers had for spending money. I did not discourage them, but I did wonder what their main motive was. Putting on my naturalist's hat, it soon became clear that displaying to each other is a major human activity. We need to understand what part display has played in our booming population and economy; and then work out how we are to manage it, as well as the population and economy.

---

[89] Lewis (1989) gives a readable, if partly fictionalised, account of bond trading.

# Chapter 12

## Biophilia & Lekking, Health & Birth Control

Biophilia?[340] Why do we feel an urge to rescue a mouse from a swimming pool? Is it because we see a living thing struggling against inanimate nature, in this case water? Yet we would set a trap in the larder. When teaching 14–16 year olds, I used to ask the class to imagine that they were driving along a country road and, at various times, their car hit and injured an earthworm, a child, a fly, a dog, a frog, a mouse, a small bird and a lizard. I asked them to write these kinds of animal in two lists: one list of the injured creatures they would stop the car and go back to help, and the other list of those they would not. If I remember correctly, and I regret that I did not keep records of the results over several years, all pupils would stop for a child or a dog, and none for a worm or a fly. Most girls wrote that they would stop for a mouse or a bird, and boys generally would not. I then pointed out that their choices roughly reflected the time since we separated evolutionarily from that animal, as measured by the proportion of DNA that we shared with it.[90]

My pupils were expressing degrees of biophilia, which is the affinity between all living things; and the closer we are genetically related (that is, the greater is the percentage of DNA that we share), the stronger the bond between us. Displaying that we are willing to recognise and act on this bond is an important element of human behaviour, and a major factor contributing to the expansion of the human population. Pity and caring have become two of the greatest human virtues.

James Lovelock, doyen of the Gaia theory of the Earth, reckons that, at present rates of energy consumption and material processing, the Earth would support a population of 1

---

[90] Dawkins (2004: p. 468 *et seq.*) gives a good explanation of how this proportion is worked out.

billion people.[91] According to an official United Nations announcement, our population passed the 7 billion mark on 31 October 2011, so we need 6 billion more deaths than births quite quickly.

The obscenity of including the word 'need' in the previous sentence is deeply offensive to civilised and feeling people. It defies and defiles all that we hold sacred. Even so, we need to break the greatest taboo of all: we need to question the sanctity of human life. Our DNA-driven certainty convinces us that all human life, however long or brief, however disabled or criminal, must be preserved at all cost. Now we are beginning to understand the consequences of continuing with that certainty. Those consequences are not just overpopulation; many of them are personal.

In 2011 I knew of an 85-year-old man dying of metastatic prostate cancer. He got out of bed, fell over and hit his head, had a 7-hour brain surgery; this failed, so he had another one. Before he died, he remained for several months in a drugged semi-coma with occasional moments of partial lucidity at a cost that nearly forced his wife to sell their home. In another case, a relative of mine became physically incapable, was admitted to hospital, fitted with feeding tubes he repeatedly pulled out and cast aside over a period of six weeks in intensive care. Eventually he was allowed to die, which he did a few months before his 100th birthday. Many who read this book will have similar experiences or hear of them through the media: newly born babies who need extensive and dramatic surgery even though their subsequent prognosis was not good,[92] and life-sentenced criminals who fall terminally ill, all receive the best attention that modern medicine can devise. Similar examples are

[91] James Lovelock in the BBC programme titled *Beautiful Minds*, 13 April 2010. Of course, the energy consumed would have to be renewable, not fossil, and materials would have to be grown or recycled rather than mined.

[92] My own grandson was born with the great arteries of his heart reversed and needed just such massive surgery to survive. He is now (2015) a rollicking 12 year-old with good prognosis. Only you who read these lines will understand what it cost me emotionally to write them.

legion. What medical resources are diverted to these cases from people in productive midlife who also needed them?

**********

What is going on here? Is it just biophilia, or are there other underlying trends or motivations? We have already looked at and understood the DNA imperative and how it evolved, and how our cooperative nature also evolved to protect copies of that DNA in other people. Doctors' Hippocratic Oath is a distillate of this drive. First inscribed in the 5th Century BCE, the oath may have had Babylonian predecessors more than a millennium before that. Succinctly, it requires a doctor to do his best for every patient, whatever their condition or nature. Modern doctors are not required to swear the oath, though it is explicit in their training. A surgeon once told me that he had broken it in a moment of rage by saying to a motor car accident patient, 'I have never said this to anyone before, and I hope I shall never have occasion to do so again: I am very sorry to tell you that you will make a full recovery, which is more than can be said for the little girl who, in your drunken spree, you put in a wheelchair for life and orphaned.'

But there are at least two other, evolutionarily more recent, influences involved. First and most obviously, there is the health industry. Millions of people depend for their jobs on medicine and caring: doctors, nurses, hospitals, ambulance drivers, the pharmaceutical industry, residential care for older people, to name a few, and they all need patients. And most of us are prepared to support this industry with our taxes – at least in countries in which there is a state-financed health service. Why most of us feel this way is due to biophilia, and it expresses itself in caring. But there is another level at which caring becomes a social token. It is called *lekking*.

## Lekking

Dawn mist clinging to the edge of a dripping pine forest; it is spring and the day will be warm. Strange calls come from the edge of the forest: a crowing-hiss and *rookoo*, and the occasional clatter of wings.[341] A dozen male black grouse are strutting with wings drooping and tail fan erect to show brilliant white bottoms;[93] healthy red combs blaze over bright eyes. Each bird has its territory adjoining those of others – the more successful birds nearer the centre. From time to time, two males approach each other at their common boundary, bow and hiss and *rookoo* to each other, flashing white underwings.

A serious fight is rare. Males lunge at each other, aiming for the red combs; but one soon retreats into his territory – a sign of submission. If the bird that retreated held a territory further from the centre of the group, that was an end of the dispute. If he held one nearer the centre, the victor would pursue him and drive him out, taking over his territory. Then the dead-leaf patterned females arrive and sit in a critical ring, watching what is going on and assessing the males. One by one they enter the territories, usually making for the most centrally placed where they will mate. The arena is called a *lek*, and the whole ritual *lekking*.

How much of our behaviour is lekking? Natural history reveals some interesting patterns. Notice that in the black grouse there are two distinct phases. First the males display to each other to establish a hierarchy of dominance, the most dominant individual claiming the most central territory in the lek. This has obvious survival value in that the birds are not only making a lot of noise to subdue their rivals and attract the females, but also aggression and sex occupies their attention. Foxes and wild cats know this and take advantage of it: male black grouse on the edge of a lek are the first to be killed. How often do you see men courteously jostling for the 'best' seat in a room? I bet that it does not have its back to a window or a door. What do women looking for a mate make of seating plans?

The second element of lekking is to impress the females, and signals of health,[342] such as clean white bottoms, brilliant red combs, glossy plumage, bright eyes and vigorous behaviour, are

---

[93] Ground-feeding birds have a high risk of picking up intestinal parasites from the droppings of others; accordingly, most males erect their tails when displaying to females to show that their cloacas are clean. This signals that the individual has few parasites, so he must be very good at the business of life, that is, he has 'good' genes.

all important. There is an overlap here, because other males will pick up these signals too, and feel inferior or superior accordingly.

It is the females that have imposed these conditions on the males by their choice of the characteristics they prefer in those with whom they will mate. In some gallinaceous birds, such as the domestic chicken and pheasant, males draw special attention to their tails. The peacock has taken the matter to an extreme: the tail covert feathers are hugely elongated and adorned with brilliant eyes, and are erected into a huge fan by the true tail beneath.

In an experiment, with long-tailed widowbirds in East Africa, researchers artificially lengthened the tails of some males and found that they were more successful in mating than males with normal tails, even though shorter-tailed males held onto their territories.[343] It seems that the tails of widowbirds and peacocks have taken on lives of their own analogous to idenes. Finally, it is worth remembering that males carry not only the genes responsible for long or many-eyed tails, but also those that will induce their daughters to measure them; just as peahens carry the genes for many-eyed tails that will express themselves in their male offspring. Our material world is full of peacocks' tails: look through any list of luxury goods. In the 1960s of my college days, sports cars were called 'bird bait', and driving them served the two black grouse purposes of impressing rival boys and weakening girls' resolutions.

Our classification of what we transmit to future generations (genes, thenes and idenes) helps us to another layer of understanding. When we are younger we display the effects of our genes: young bodies are more beautiful than older ones that show life's wear, and flaunted with effect. After all, the young have generally not had time to achieve much that would be worth showing off; the best they can do is to boast of their genes. As we get older, thenes become more important. On entering a room or joining a group, young men often put their mobile phones and car keys on a conspicuous table. This serves two purposes: it shows off the cost of their communications and transport, and also avoids spoiling the line of their clothes with ugly bulges. Money is more difficult to display because doing so is in itself bad taste, therefore money is converted into thenes that can be exhibited less flagrantly.

Displays of practical skill, such as reversing a car into a small parking slot in one precise movement are very important when a group of young men gather together. I wonder how many youths are caught speeding when driving with other men, compared with those that are alone or with a girl. I guess more. Wit is an example of bare idenes because it can hold several people's attention at once, so is an effective lekking device.

As young people get nearer to reproduction, their lekking tactics change: they are more concerned with showing what good parents they would make. When we give conspicuously to charity or call out, 'Do be careful!', we are calling *rookoo*, that is, we are sending out two messages that translate into 'biospeak' for the opposite sex as, 'Look what a caring person I am, I would look after your children as carefully as this if you mated with me.' However, this biospeak translates to the same sex as, 'I am more caring than you so more likely to pull a mate; step aside.'

Political platforms make excellent middens from which to crow this single message with two meanings. Indeed, since the terrorist outrages in the USA in September 2001, being seen to care for public safety has become another industry that is closely linked to medicine. Some airport security rituals are more signals of the state caring than effective measures against attack. A politician who did not toe this line – or, worse, spoke against it – would have no chance in an election. Newsreaders on radio or television and other media people adjust their faces and tone of voice appropriately when announcing a disaster, loss of human life, rescue, or discovery of a new medical advance. And they do this because what they say strikes a chord within every human breast. That chord is biophilia – the common identity that all living things share according to the percentage of common DNA.

With the immediacy of modern communications, we now extend the 'care for' elements of our lekking display to embrace all mankind, even in far-off lands; and we do this with aid to poorer people. Should we not also reach out with our love and pity in another dimension: the dimension of time as well as of space, and so include generations yet to be born? As explained in earlier chapters, DNA cannot predict; this is why it does not encourage us to consider future generations.

Though our biophilia and pity are boundless, the practical resources we can call on to turn these feelings into actions are limited – because the Earth is finite. How, then, shall we apportion what we can do? It is easy to reach for a chequebook or click a computer button when we are moved by the latest tragedy on television, and doing so makes us feel better. Good feelings make us happy, and happiness is what moral people should be aiming at, if the Constitution of the United States is to be believed.[94] So what is happiness? Even if it were no more than the release of dopamine in brain cells, neither society nor government have any business trying to promote it. All they can do is to lessen unhappiness by reducing the practical causes of it.

If, as the philosopher Jeremy Bentham (1748–1832) claimed, a truly moral action is one that achieves the greatest good for the greatest number of people, and that 'good' in this case means 'happiness', surely it is moral to allow people the pleasure of dopamine release when they help to relieve suffering whenever they hear of it? Again, I come back to that word 'more' or 'the greatest number'.

Bentham makes no mention of the dimension of time, indeed he rails against consideration of it, believing that it was absurd to torment the living under pretence of promoting the happiness of those who are not born, and may never be born.[344] Even so I repeat my contention that we ought to. There are 134 million children born every year, and 56 million of us die. That is 21 born and 9 die – a gain of 12 people – every 5 seconds. With

[94] The Constitution of the United States includes the phrase '... unalienable Rights [to] ... the pursuit of Happiness.' Encyclopaedia Britannica (1991: v. 6, p. 283).

this in mind, now what do we mean by Bentham's 'greatest number'? If we do take the future into account, pursuing happiness now condemns all children alive today to a terrible future.

How are we to practically resolve this conflict between present and future, between hearts and minds? We feel the power of DNA's imperious demand that we care for copies of it in other living things, yet our minds tell us that doing so will so overcrowd the Earth, that we will make it no fit place to live in for those who come after us. Watching people die when we could help them to stay alive stirs up appalling feelings of guilt. We can avoid this by simply preventing babies from being born. That is not killing people, so we do not feel so badly about it.

But there is an evolutionary problem: reducing our birth rate reduces variation, and variation is the raw material on which evolution works. This is even more important in a genetically impoverished species such as ours. Even though the conspicuous parts of our evolution have become more behavioural than genetic, we still need new variation in our DNA to combat the counter-evolution of our disease-causing parasites.

As we have seen, our bodies are temporary custodians of the genetic germ line that we each inherited from our parents, and which we can pass on to our children. On average, in the course of our lifetimes, each one of us has one mutation in the germ line we carry.[345] Such mutations do not usually appear in our children because, when they recombine into pairs at fertilisation, they join with a normal healthy gene and the defect is masked – it is said to be *recessive* to the non-mutated gene's dominance. A mutated gene would only have an effect when it happened to meet another copy of itself, until its dominance evolved.

The advantage of a mutation in the germ line of an individual is that it causes the genome to be *different*, and a different genome produces different proteins. Parasitic microbes do not recognise the different proteins as belonging to the species of host they have evolved to parasitise, so they are less likely to infect it. The genetic material in the parasite also mutates, and by chance can allow the parasite to recognise the newly formulated protein combination as belonging to a suitable host.[95]

Most mutations are minute changes in the genome. If one were big enough to show its effect in the child, that effect would almost certainly be lethal, and the embryo would abort at an early stage in its development. Even recessive genes may make tiny alterations to the whole internal chemical environment of the body, and that can trigger spontaneous abortion. Medical intervention can overcome this tendency and give parents the happiness of a seemingly healthy child. But the mutation persists in the child's germ line adding yet another to those accumulating in the whole human genome. We selfishly regard our present happiness as more important than the child's agony when it grows up and one of its sex cells meets a similar mutation in its partner. In sum, medicine has severely reduced the way natural selection eliminates harmful mutations. The consequences are clear: we are heading rapidly towards a situation when most of human effort will have to be devoted to building, maintaining and running hospitals.[96] Present (summer 2015) concerns about the UK National Health Service (NHS) and care of older people are certainly the first signs of impending overload. I discuss this in its wider context in the next chapter.

Though increasing numbers of women desperately want effective methods of contraception, a century of effort in this direction has not reduced our birth rate, nor has it reduced its rate of increase. The main reason is simple – there are more women. Indeed there are nearly 1 billion prepubescent girls, and

---

[95] Indeed, the process of sex – halving the genome and restoring the full number by fertilisation – seems to have evolved as defence against attack by parasites (Hamilton, 2001: ch. 12).

[96] Hamilton (2001: ch. 12). *The Hospitals Are Coming* is a clear scientific statement. The section also contains a most perceptive and beautiful piece of writing that deals with the reality of death: the deaths of two of Hamilton's brothers and his dog (pp. 477–483).

all of them will exercise their right to have as many children as they want.

If it were practically possible to limit the number of children each woman has, the process is likely to take too long for our population to fall to truly sustainable levels before civil strife intervenes. Thus we are forced to consider other ways of reducing our population. The sooner we face the stark fact that we are going to have to take human life deliberately, the less we will have to.

# Chapter 13

## Taking Human Life & Species, Aims & Morality

Birth control is preventing new life, but killing is entirely different. There are two backgrounds to deliberately taking life: passive and active, and they blur together because deciding not to do something is an active choice. We will look at the less contentious end of the scale first. The mere fact that we are a highly social species lays a duty of care to others on each one of us. Which is to say that every right we claim to be helped when in need, carries with it the duty to help needy others. This balance works well, and has contributed to our success as a species. Now that our population has become unsustainable, and our ability to think has freed us from the bonds of our genetic inheritance, we have a duty to question the nature of the right and the duty to care.

We will look first at some possible courses of action that, at first sight, seem passive. For much of this century, the UK National Health Service (NHS) has come under increasing pressure as costs and numbers of patients rise;[346] there will almost certainly come a time when the state will have to stop caring for certain categories of patient. Such a proposal would surely be received with outrage by much of the communications and social media, and, under their emotional pleas, the electorate also. That protest can be balanced by persuasion through rational argument. The general outrage would be due to feelings and precedents which some consider to be an essential part of what we mean by being human.[347] As we have seen, these feelings are driven by DNA, and the precedents derived from it. Even so, I shall discuss four categories of patient from whom we should consider withholding medical intervention.

First, and perhaps least contentiously, we should consider allowing people to die when it is their considered wish. It is certainly mine. My right to life (which is a human construct and not a law of nature to be discovered) does not carry with it a right to expect someone else to devote their life to prolonging mine. All I would ask is relief from terminal pain. The number of people who voluntarily travel to countries where assisted suicide is allowed, and pay much money to do so, shows that there is a significant tide of opinion in favour of ending one's life before the indignities of incapacity set in. More drastically, it is likely that we will have to decide not to give medical help to people over the age of, say, 75.[97] Of course this would depend on the contribution they made to the welfare of others, even if that be no more than company for someone else.

Another category of our population that might be considered for withholding medical treatment, are those suffering from self-inflicted injury or disease. Each one of us has a duty to care for their own body;[348] those who do not do so have no moral claim on the state to help them prolong their lives. It would certainly reduce National Health Service (NHS) costs and physical risks to staff, if drunks seeking help were told to go away and come back when they were sober. Obesity is a rising problem throughout the overdeveloped world, and it can be quantified objectively. The body mass index (BMI) is easily calculated by dividing one's weight in kilograms by one's height in metres, dividing that by 1.75, and fitting that figure to the scale: *underweight* (under 18.5), *normal weight* (18.5–25), *overweight* (25–30), *obese* (over 30). A simpler test is to kiss your knees.

Obesity not only increases the risks of heart disease, stroke, diabetes and cancer, which impose further strains on health services, but also sends out a signal of contempt for people who do not have enough to eat. Again, as population pressures rise, the NHS may well have to tell obese, or even overweight, people to come back for treatment when their BMI is healthy. It might help to prevent this if obese people were charged according to their BMI on public transport, though the practical problems would probably make it too expensive to enforce.

The next category of people from whom we should think about withholding medical intervention are babies, and that

[97] I was born in 1938.

takes us into a minefield of emotional reaction. The DNA imperative to 'save' (= prolong) the life of every child is so deeply ingrained in our genetic being that we call it sacred. Indeed, we take it to be nature's or a god's absolute law. Our minds are not so sure, because now we understand that it is merely DNA pulling the strings of our behaviour to replicate itself. But DNA is blind to the future, it cannot predict, whereas our minds can. We have no choice other than to face the probability that, when our population reaches a certain level, the state will have to stop paying for pre- and postnatal medical care before the age of, say, two years. The evolution of our birth process is far from coping perfectly with our huge skulls and our habit of walking upright, let alone our modern life-styles. As I explained in the previous chapter, every baby's life we prolong by medical intervention to the point when they can breed, accumulates deleterious genes in the human genome and compounds their impact on future generations. At least let us remove this subject from the list of taboos and debate it.

Withholding state-funded medicine anywhere raises the question of whether, in the interests of justice and fairness, government should curtail individuals' freedom to pay for such medical intervention. We have seen how justice and fairness evolved as social devices to promote group cohesion and hence competitiveness with other groups, and were not handed down as divine creeds. Whatever their origin, does that same justice and fairness extend to today's children, who will have to pay for the care and treatment of increasing numbers of lifestyle-inflicted disabilities, and genetically malformed and merely senile hospital cases? At least withdrawing medical care from any category of people is reverting to the state inaction that existed before the NHS was set up in the UK in 1946.

I suggest that the Hippocratic Oath be rewritten so that doctors can decide case by case, and be immune from prosecutions, unless evidence points to malevolence. If individuals or their families dissent from their doctor's decision, there is the time-honoured solution of courts of law, which, in emergency cases, could be convened in hours. Evidence presented in such disputes should include estimates of how much tobacco, drugs or drink, the patient has consumed, as well as their BMI.

Another category of passive population control is to withdraw medical aid, or aid of any sort, to populations that are still increasing. This would result in massive migrations, and I discuss it in its wider context in the next chapter. Although withdrawing state funding in certain medical circumstances or as overseas aid does not involve killing people directly, some may argue, as I suggest above, that deliberate inaction is a form of action. If that is a valid proposition, then it applies equally to the thesis of this book – passively not doing something about overpopulation is an active decision.

The boundary between passivity and activity is further blurred in the matter of suicide. In September 2015, the UK Parliament debated a proposal that assisting a person to die, if they had less than six months to live, should no longer be a crime. 74% of Members of Parliament (MPs) voted against the bill. The fact that 72% had voted against the same proposal in 1997 suggests that both sets of MPs were reading from the same script, rather than thinking.[349] To amend these laws must surely be a high priority for all governments.

Present objections to taking one's own life are based on superstition and muddled religious ideas of a soul implanted by a god. As we have seen, this basic tenet of humanism is founded on feelings that have their roots in DNA's drive to replicate itself. Modern attitudes are changing slowly: though there is a clear distinction between what people think or feel in private, and what they would be happy for others to know about their attitude.

There is always the risk that old or infirm people would be vulnerable to pressure to end their lives from family, friends, beneficiaries, insurers or even the state. Clearly anyone convicted of applying such pressure would be breaking the law and be subject to discouraging penalties. There will always be individuals who obstinately cling to life even though doing so risks impoverishing their families. There will also be families who insist that living, brain-damaged members of it must be kept alive at all cost – especially cost to the state.

Those who have committed shameful acts should be given reasonable opportunity to end their lives, if they so wish. Suicide is an act of enormous courage, which may be why it was, and in places still is, considered an honourable solution, and going some way, where relevant, towards expiation. There is an urgent need to make suicide legal, respectable and eventually admirable, as it was in Roman times. There is no rational philosophical reason why not. At least let us start with the matter of suicide generally being open to thoughtful discussion, and all knowledge being freely available.

**********

Before moving on to coercion, we need to discuss the concept of its counterpart: freedom. Its fundamentals are well described in JS Mill's *On Liberty* and in Isaiah Berlin's *Liberty*. Both authors struggled with contradictions in the ways we exercise freedom, but these difficulties largely melt away when the concept is seen through a naturalist's eyes. As described in the earlier chapters of this book, all activities of life, even among its precursors, are reducible to the fundamental choice: compete or cooperate. So it is with the idea of freedom.

The idea that every individual is free to act as they wish, so long as they do not impede a like freedom in others, is hard to improve on. Unfortunately, it is frequently invoked without its corollary, which is 'so long as the act is at the actor's own risk and peril'. This last proviso is of course indispensable.[350] Not only does it give impulsive people pause to think about what they are preparing to do, but it is also fundamental to the mechanism of evolution, which is the dominating force in the formation and development of ideas as well as bodies.

'A like freedom in others.' Who are these others? Again, I ask whether they include children who will be adults when the full ferocity of the population tsunami breaks upon us. Despite Bentham's cynicism, I also include generations to be born, as I explain towards the end of this chapter. The concept of freedom cannot be defined in a sentence or two; its true meaning can only become clear by using it in many contexts. Our present difficulty is that these contexts are changing rapidly as our understanding of what it really is to be human grows.

**********

Now we come to deliberately taking the life of another person, and the slightly less contentious starting point is probably ending pregnancy. Because it is difficult to establish paternity in so many cases, the onus of reproduction limitation must be upon women – as it is in most cases anyway. The replacement birth rate is about 2.1 children per woman,[351] so a limit of two children per woman would steadily bring the population down while still supplying some genetic variation. It would be easy to pass such a law, but practically impossible to enforce it.

Before discussing the issue further, we need to look at more of the biological implications of compulsory birth control – implications that are perhaps more raw than many who read this book can comfortably stomach. First, we would have to decide whether to count a pregnancy with two living siblings as an infringement, or a third pregnancy *per se*, regardless of whether the previous two resulted in living children. The latter is biologically preferable for the genetic reasons already discussed. Practically, any woman four or more months into her third pregnancy could be counted as criminal, regardless of whether she has living children, and even if she is a victim of rape because she has had time in which to arrange an abortion, which should be legally financed and controlled.

Possible sanctions against a third pregnancy are compulsory termination and sterilisation. The sheer logistics of enforcing it in poor countries would be tantamount to war with 'troops on the ground'. Again, we must ask ourselves whether such action now is better or worse than leaving to our children the decisions how to sort out the far greater mayhem that would inevitably follow our inaction. I repeat that I am appalled by these thoughts, which emerge as a result of dispassionate thinking; but not pursuing them to their practical conclusions would be an act of intellectual cowardice.

The idea of forcible abortion and sterilisation of women in poor countries is indissolubly linked with food, medical and monetary aid, not to mention weapons by which scarcely democratic governments hold power. Without that intervention, people would remain poor, and many would die of disease and starvation, as they did before aid was given. At that time, the death rate kept balance with the birth rate, and the human population survived just as wild animals do, by maintaining a dynamic tension, which is a balance, with the resources available. Providing medical help and food ensured that many children did not die, and populations rose to levels that are sustainable only by more and more aid.

The present situation in the developing world is that we (all of us) must make a choice between three courses of action: (1) we go on providing food and medical aid to the existing population plus five more babies every two seconds; or (2) we (the richer of us) accept the surplus people into our own countries and share our living space as well as our social services with them; or (3) we (all of us) inure ourselves to the harrowing spectacle of literally millions of people dying of starvation. No wonder politicians and media persons prefer to evade the decision and postpone it to future generations. But, for our children's sake, we have to decide, even though we will be divided in our choices, and also *by* them. And this will mean war of one kind or another.

**********

Terminating a pregnancy is deliberately taking human life and leads our argument into even more difficult areas. Before tackling the issue of capital punishment, there is a wider and deeper group of concepts that should be part of the discussion.

An unarmed man escapes from custody in an asylum into a town and starts hitting and biting people. The police restrain him and cart him back to custody. A male chimpanzee escapes from custody in a zoo into a town and starts hitting and biting people. The police have no hesitation in shooting him dead. Why the difference? Because we are two separate genetic species.

The difficult word here is 'species', and it needs discussion. There are several meanings of the word,[352] which can be named by prefixes: a *biospecies* is a group of individuals that share in a common gene pool, and we can test whether they do by attempting to breed them.[98] Obviously we cannot usually apply such a test between an individual and its grandparents; equally obviously they are of the same biospecies. Likewise, it would be perverse to regard people who had lived in the 18th Century as not belonging to *Homo sapiens*.

As one goes further back through the generations, it is convenient to think of *paleospecies*, which we use to describe groups of fossils that share distinct features. Similarly, '*museospecies*' are groups of specimens in museum drawers that can be distinguished from other such groups – often by apparently trivial features. Both categories being dead, we cannot test whether either is a true biospecies.

Then there are certain groups of living plants or animals that share so many physical and behavioural characteristics that it is convenient to describe them as a single species; but they have given up sexual reproduction, so we cannot apply the breeding test to them either; we call them *agamospecies*.

Some sexually reproducing plants and animals are likewise so morphologically similar that they cannot easily be distinguished by structure, yet they inhabit different ecological niches and do not interbreed, even though they can under artificial conditions; we can call them *ecospecies*.

Giraffes are an interesting group of mammals that occupy much the same type of habitat, but the populations that inhabit distinct regions of Africa show marked differences. Some authorities describe them as races or subspecies because they can interbreed freely in captivity. However, given a free choice, giraffes prefer to breed with their own, what we have hitherto called, race or subspecies; DNA analysis has supported this.[353]

---

[98] The concept of a biospecies is blurred by grades of hybridisation, such as donkey × horse (two distinct species) = mule which is sterile.

Perhaps we can call the assemblage of new giraffe species '*patriospecies*' (Latin *patria* = 'homeland').

Then there are animals that appear identical, inhabit the same area and much the same sort of habitat and are interfertile, which is to say that their sperm will fertilise their eggs *in vitro*, but the natural behaviour of the animals normally prevents them from interbreeding, for example, they mate at different seasons or times of day. They can be called *ethospecies*, from 'ethology' the study of behaviour.

We consider the concept of biospecies as fundamental because our feelings and mental models are still ordered largely by DNA. Which definition of species we use depends only on how we want to use it. From the evidence of interracial breeding alone, it is clear that *Homo sapiens* is a true biospecies. Yet we have become successful by adapting our behaviour, therefore we can think if ourselves as also being an ethospecies.

When we look at the fossil record, we can see how one species evolved into another, sometimes a single species evolving into several new ones. This is a process called *speciation*, and it is often preceded by the parent species becoming common. Speciation makes obvious sense, because, when too many individuals are competing for the same limited resource, selection favours those that diversify.

The fossil record leaves little evidence of behaviour, but that should not discount its effectiveness in speciation, both in the past and today among living species. Indeed, I suspect that all speciation begins with a tiny alteration in behaviour, like a caterpillar eating a previously untried species of plant. If such a change were advantageous, positive selection would favour it because it reduced competition between it and those individuals that continued to feed on the traditional larval host plant. Eventually, if the reduced competition continued to be selected for, the new behaviour would have become incorporated into the animal's genome, and the species' anatomy and physiology would be changed.

It is clear from differences in structure and physiology between different human races that we began to speciate genetically. Bear in mind that, because we all went through one, and most of us two, evolutionary bottlenecks, we have little genetic diversity when compared with our nearly related species. Thus we did not diversify much before races began to travel, met each other and mixed their genes, and this effectively prevented biospeciation. Genetic evolution can only take place in a population that is isolated from such a gene flow. The only way *Homo sapiens* could speciate genetically now would be by populations of people cutting themselves off reproductively from the rest of the world for hundreds of generations.

Behavioural evolution is another matter: it can take place in seconds, it is largely under our voluntary control, and it is entirely independent of our racial origins, though not our upbringing. Since we object to killing members of our own species more than those of others, would it help to overcome this barrier if we regarded some of us, who behaved badly, as members of a different species? No, it would not help, at least to start with, because each of us carries with us such a heavy load of stereotyped genetic baggage: we are driven by mere chemical DNA to feel that we should protect copies of it in other members of our biospecies.

Even so, we have become so abundant, and have so little time left, that the need for us to *ethospeciate* is pressing. I repeat, we can do this only by rational persuasion. Could we imagine a time in which each of us individually decided on our species? For example, by obeying the law, an individual would remain a member of *Homo sapiens*, whereas by breaking it they would clearly signal that they had chosen to leave that species and join '*Homo deviens*', or some other more aptly named species. To help understand etho-speciation in humans, one more suite of ideas needs to be brought into the argument, and it will seem at first glance to be insufferably abstruse. It is, why are we here?

'Why?' means both 'How did we arrive here?' and 'For what purpose are we here?' Much of this book describes the former, but here I mean the latter. Before discussing this shorter but deeper topic, we need to be clear about what we mean by 'we'.

Does 'we' mean only the living? Does 'living' mean from the moment of conception? After 24 weeks? After birth? There are plenty of distractions here from the main argument. Does 'we' include the dead? If not, then what is our purpose in making a will or other testamentary disposition? We, the living, have a contract with the dead to do their will.[354] But we can hardly be expected to honour our ancestors if they have squandered our family fortunes. Nor should we expect our children and grandchildren to respect our wills and wishes, if we have so despoiled the Earth that it is practically uninhabitable by them.

Our children are our future. They are the vehicles carrying our genes and idenes forward (assisted or impeded by thenes), and hand them on – individually inevitably diluted – to succeeding generations. That is why I include them in 'we'; they are our immortality. I know of no testable evidence to support the idea that each one of us will arise as a recognisable individual in some afterlife.

Pursuing the idea of immortality a bit further, the only unique genes we pass on to future generations are the one or two that we generated individually by mutation – and all but a tiny proportion of these are deleterious. All the other millions of genes we bequeath to our children we share with other human beings. Our genetic immortality is as much in caring for other people as in our children.

**********

William Shakespeare is immortal because of his writing, not because of his children; Antonio Stradivari because of the violins he made; Albert Einstein's immortality is an equation, and Charles Darwin's another idea.[99] Few of us will achieve such individual immortality as these four artistic or scientific giants, but we all can contribute something. The heroes I cite have been supported and helped to accomplish what they did by

[99] The difference between them is that, had the first two never lived, we would not know of *King Lear* or a 'Strad', but someone else would certainly have discovered $E = mc^2$ and the idea of natural selection, as indeed did Alfred Wallace. Creation and discovery is the difference between art and science.

those who supplied them with food and clothes, light and drains. This is the immortality we more lowly humans can claim and achieve. Every little act of kindness to other people contributes to the welfare, and hence immortality, of all. Certainly some jobs are paid more than others, but that is because fewer people can do them well: their wage is a reflection of the market forces of supply and demand, not necessarily some imaginary scale of usefulness or importance or, worse, family. This whole spectrum of contribution to both collective and individual immortality is behavioural, not genetic. In short, 'we', in this case, means all human beings who contribute to the welfare of others, stretching as far back as we care to think and forward certainly into the 'foreseeable' future.

**********

There are three vital differences between genes and idenes. The first is that genes use up vast resources of the Earth to propagate themselves, while idenes do not. This means that there is a limit to the biomass of human genes the Earth can support; but, because idenes are so undemanding of the Earth's resources, there is practically no limit to the number of them the human mind and its machines can store and propagate.

The second difference is that genetic evolution works on a timescale of generations, in our case, in units of 20–40 years for each mutation to appear, and then more generations for its effect to manifest itself in the physical form or behaviour of new individuals. In contrast, idenes can evolve and show their effect in seconds. The death of Diana Princess of Wales affected only a handful of people; but *the news of it* (an idene) changed hundreds of millions the moment they heard it.

The third difference is that genes cannot foresee. They are entirely driven by hindsight: evolution works only by what *is*, never by what shall be. Thus, 'what shall be' in genetic evolution is merely a human estimate, never a genetic intention. On the other hand, human thinking, which consists of idenes, *can* predict: thinking can assess frequency and probability. It has set us free from the tyranny of mere chemicals,[355] and in so doing, it has become our future.

It follows from the differences between genes and idenes that it is worth trying to preserve the latter, and to use genes and thenes for this purpose. Hence preservation of information, which includes thinking, idenes and knowledge, could and, in my view, should become our collective aim. This would unite us with a third attribute that stands beside sharing in one common gene pool and the biosphere of planet Earth. More than that, it would be a behavioural attribute we can alter by our own free wills.

Even so, our idenes need the human mind to propagate them, therefore those who have children, and those who supply the necessities of reasonably civilised life, are contributing to the survival of idenes, and their work should not be disparaged. In this way, the DNA imperative, which is no more than chemicals trying to make more of themselves, can be harnessed to serve the wider aim of idene survival. Note that I have nothing to say about what those idenes shall be. I would like to think that Darwinian evolution will operate among them, and arrange for our chemical bodies and military thenes to control idenes that might damage the appearance of others in future generations.

**********

In Chapter 9 we established a natural image of what it really is to be an individual, and from that image arose a satisfactory understanding of immortality. Both the concept of the individual and the concept of immortality are streams flowing through time: streams of materials and streams of information. (From now on, I shall use the word 'information' to include behaviour, thinking, idenes, knowledge and information contained in all entities, whether it be understood or not.)

The main difference between individuality and immortality is that we define an individual as those parts of the two streams that interact between the individual's moment of conception and its moment of death. During that time, which we call a life, a constantly changing side stream of information is managing a constantly changing stream of materials. Thus the individual is a device on which selection can operate to alter, minutely, the vast universal river of information that flows through time. Giving special recognition to the idea of the individual was an important step in our evolving sociality; therefore it is a human construct. If we see it as that, and discard misleading words such as 'artificial' and 'natural' and 'soul', we are clearly responsible for consciously adapting, in the light of our modern understanding, how we regard 'the individual'.

This preamble prepares the way for deciding on the purpose of our being here, and I suggest that it be to safeguard, enhance and propagate information. It is an achievable aim that does not use up the world's resources excessively. I repeat: it also gives a third leg, and hence stability, to the stool that supports the unity of all human beings: we share the planet Earth, we share a common pool of genes, and now we can share the aim of safeguarding, enhancing and propagating information.

Only a small change of wording turns this aim into a new definition of morality:

> 'Any action that tends to preserve, propagate or increase information in entities other than the actor may be regarded as moral, even if that information be the same as in the actor.'

If we can agree on this, then it could restrain our unthinking desire to 'save' every life, however criminal or diseased. And this brings us back to practicalities.

**********

We have divided up our own biospecies into nation states, and we did so originally on a broadly genetic basis: races and peoples recognised structural and behavioural patterns that they shared because of their recent common genetic ancestry. Our evolutionary history of competition and predation endowed us with suspicion and fear of strangers. As political systems evolved, people aggregated into nation states, often to defend themselves from occupation by people with different languages and customs.

In some cases where nationhood was imposed by conquerors, they got it wrong. For example, in the former colonial and protectorate countries of West Africa the national boundaries generally radiate along watersheds from the centre of the landmass to the sea. This is because the colonists gained access to the interior by boats on the great rivers and traded with people on the banks, often failing to perceive that people on one bank were different from those on the other. When the colonists took over the administration of the land, their economic boundaries became the watersheds. Unfortunately, West African people had found that rivers were a greater obstacle to travel than ridges, so they had made rivers their tribal boundaries. Thus, when colonial administrations gave way to nations, tribes found themselves divided and, worse than that, forced to share political decision with traditional enemies across the river.[356]

Under the European, Soviet and Ottoman empires, ethnic divisions were often overridden and people forcibly moved around. Stalin's translocation of people from North Ossetia southwards over the Caucasus Mountains into Georgia led to war in 2008 and the occupation of Georgian territory by Russians. The almost permanent conflict in the Middle and Near East has many of its roots in the Anglo-French carve-up of the Ottoman Empire. The ambition of the so-called Islamic State to erase these divisions could be laudable, were it not for their methods and other aims.

Yet still political maps of the world dictate what laws we shall obey and how we treat other peoples. Though all my ancestors have been English for at least four generations, I find a much stronger affinity between law-abiding people throughout the world than with criminal English. Dare I add that I can refine the category of my personal affinity by adding the word 'reading'? Is it not time for us to think carefully about what criteria we choose to define a nation? In this age of instant global communication of ideas, have we not outgrown mere animal territory? Certainly multinational business empires pay scant regard to national boundaries, except insofar as differential tax and regulatory regimes affect them.

Though several authors claim that territory has been hugely important in the evolution of loyalty,[100] I suspect that politicians and religious leaders have manipulated much of this 'sense of place'. For some people, shared beliefs bind them together. Beliefs are idenes and they occupy a large proportion of storage space in minds, files and print. When loyalty to a belief is not shared with neighbours, and has its roots far away from where the adherent lives – places such as Jerusalem, Rome or Mecca – social tensions soon appear. Loyalty to the belief that there are no mysteries, and that all relationships between cause and effect are understandable, and that preserving information is our common aim, has no geographical or racial focus.

During the Enlightenment in the late 17th and much of the 18th Centuries, when empires blurred national boundaries and passports were unheard of, scholars travelled freely to each other's countries for meetings and discussion. Even in a time of revolution and war, the Royal Society in Britain and the *Académie des sciences* in France corresponded and exchanged representatives.[357] This intercourse extended to North America as well as other European countries, and the participants became known as the *Republic of Letters*. Members of the various contributing societies left royal families and politicians to squabble over mating opportunities and thenes and went on with their primary purpose, which was in the realm of idenes.

**********

For practical purposes, behaviour is under the control of the will, while anatomy and physiology are not, and that vital difference is the key to our future. It is also why any form of racism is so utterly repugnant. Criteria for separating people must be based on ethical not ethnic grounds, but this should not debar discussion of, and research into, possible connections between race or genotype and behaviour.

The next difficulty is to decide how we are going to introduce these ideas into our societies. There are many levels of

---

[100] Scruton (2006: p. 11) and on many other pages not mentioned in the index.

behavioural conformity: norms, customs, traditions, rules and laws. They are labelled with words we use to describe blurred bands on the spectrum of our social behaviour. They grade smoothly from one to another, and are rooted deeply in our behavioural evolution. We need to look at them before suggesting at which level it would be most effective to impose the ideas we have discussed so far in this book.

# Chapter 14

## LAWS & POLITICS, REPUTATIONS & PUNISHMENTS

However we answer it, the London question is easy – at least compared with others that will inevitably follow from it. In practice, we can reduce all decisions connected with the population tsunami to one simple collective question: 'Do we want the inevitable contraction of our population to be under the rule of law, or not?' Doing nothing is simply deferring responsibility to future generations, so is an active choice for lawlessness, and so also is the denial that law and order will break down. The recent history of North Africa and the Near East is evidence enough that this process has already started, not to mention the hordes of displaced people throughout the world.

An orderly reduction in population can be achieved only under the rule of law: there is no practical alternative. Imagining some benign, Mahdi-like dictator who would arise and lead us out of the dilemma is futile. The difficult decisions are ours individually, and we must face them with as much clarity of thought as we can muster, even though our feelings and ancestral rules scream otherwise.

Once again I shall approach Law from the point of view of a naturalist, including its recent evolution. As mentioned at the end of the previous chapter, laws seem to grade into other institutions and the less formal regulatory devices of society. It is important to understand this so that we can judge at what level to introduce the ideas we are arriving at.

For example, the idea of sharing the Earth's resources could well spread and be put into practice by a groundswell of public opinion pouring scorn on ostentatious displays of wealth, rather as it influenced the wearing of wild animals' furs.[358] Likewise, we could treat obesity as we treat smoking. But these pressures work only where public opinion matters to rich individuals. For many of them defiance is a social display, and the not-so-rich ape them. Looking at how this behaviour might have evolved throws a different light on our sociality.

All healthy mammals seek food, shelter and mates. Most of them facilitate this search with a sense of territory. We humans added stored food and other possessions, and later ideas to our 'facilities' (in the military sense). When individuals of a solitary species meet others of their kind, they usually compete with individuals of the same sex and ignore or cooperate with the opposite sex. Competition immediately creates winners and losers; individuals in either of these categories certainly recognise individuals of their own species, and defer to or threaten them according to their memories of previous encounters. This creates a ranking, even among normally solitary mammals.

In more social species, higher-ranking individuals feel no obligation to do as others expect them to. Conversely, those of lower rank must mind their manners, at least in the presence of their 'betters'. They take second place in feeding, finding shelter and fleeing from danger, and their access to mates may not be tolerated at all. In effect, a few dominant individuals lay down the law to many subordinates.

This general pattern is modified when a high-ranking individual fears to behave in a way that might unite individuals of lower rank against him. He survives longer if he is aware that they require and perceive a sense of justice in him. Such fairness need not necessarily be articulated; it can be simply felt.[359] But even this sense of justice can be subverted to selfish ends. In the course of behavioural evolution, two lower-ranking individuals soon found that, by cooperating together, they could usurp a single alpha's position. To counter this threat, alphas entered into coalitions with other higher-ranking individuals, and shared privileges with them. Cooperating to form coalitions became competitive and led to the behavioural arms race we now call politics.[360] It also led to group solidarity, the stratification of society and the emergence of classes – features that are present in all social animals. In short, animals and people cooperate to compete more effectively. The decision then resolved into: 'With whom do I cooperate?' Shared DNA was the initiating and driving force for cooperation throughout most of organic evolution, and it is still strongly active in human nepotism. Then there was the problem of how to recognise relatives.

Within family clans whose members knew one another, there was no problem about remembering who was in which interest group. However, as numbers of individuals in clans and tribes increased and it was harder to remember who was on whose side, members of interest groups evolved shorthand ways of recognising one another. They began to share personal habits, perhaps by imitating those of an influential member, and these became the social norm for that group. Ways of eating (table manners) were probably important early on, certainly preceding tables. The order in which individuals helped themselves or others to shared food has been subject to norms of etiquette since the earliest primates became social. Today, we make up our minds about people when we notice that it is the resident male who serves himself before a visiting female, or allows his children to guzzle. Then with language came the use of words: vocabulary, grammar and accent, all of which became indicators of grouping.[361] Such signals were also important in matters of security because non-conformity to a social norm indicated a stranger who might be malevolent.

Deviation from norms attracted comment and gossip within the group, and probably raised doubts about the suitability of the offender to be a member of the group. As soon as anyone breaching norms ran the risk of being expelled from the group, the norm became a rule. Games were played according to rules but the distinction was blurred when breaking certain rules attracted penalties in addition to exclusion. There is something essentially English about calling the rules of cricket 'laws'.

Because England had such a profound influence on the legal systems of the world, I shall take English Law as the central example. We have no certain knowledge of how pre-Roman occupants of England organised themselves, but it is safe to assume that they were divided into subsistence farming and eternally warring tribes. Under the four centuries of Roman rule, England was governed by a people who, though they had come to settle, looked towards Rome as the source of their inspiration and a repository of their allegiance. To Rome they paid their tribute, and it was to Rome that the legions eventually withdrew.

Roman law emerged out of military might. In the frenzy of war, it is clearly more efficient for an officer to give orders, and that they be obeyed at once, than to consult others and debate the rights and wrongs of precedent. The Romans were almost continuously at war from the founding of Rome to its eventual collapse a millennium later. Small wonder that they evolved and favoured the top-down method of Civil Law, and imposed it on their empire. Eventually that empire covered all but the Atlantic fringe of Europe west of the Rhine including England and far into Asia Minor and North Africa. Though they never conquered what is now Germany, the Romans had a profound effect on the Germanic tribes with whom they came into regular contact. These tribes adopted many Roman practices, including Civil Law, and their administrative habits were reinforced later when the Frankish kingdoms extended over much of Germany.

In the 4$^{th}$ Century CE, events in Europe initiated a cascade of consequences that led to the collapse of the Roman Empire and set the template of modern national divisions. Hunnic peoples swarmed into Eastern Europe from Asia and displaced Germanic tribes. The Ostro[Eastern]goths went south into Italy, the Visi[Western]goths to south-west France and, later, on to Spain, the Franks to northern France, the Suebi and Vandals to Spain, and the Saxons and Angles north-west to England. These vast migrations played a major part in the collapse of the Roman Empire.

It was not so much the volume of people changing places, though their genetic signatures can still be detected,[362] as tides of ideas. These ideas had their main expression in patterns of social organisation, because it was on them that the success or failure of the tribe depended. For example, the Vandals entered Spain in 409 CE and had vanished into Morocco within 20 years. In Spain, they found a culture so rich in consumable artefacts that they had little incentive to make much for themselves. They left little sign of their passing except destruction and their name in Andalucía.

Very different from Vandalism was the attitude the Angles brought to northern and eastern England, and the Saxons to southern and central areas. The Anglo-Saxons found much the same level of wealth in England as the Vandals had in Spain, but their approach to them was more constructive. These two German tribes, with some Jutes and Frisians, invaded and defeated the Romano-British occupants of England, driving their remnants onto the wetter climate and poorer soils to the west and north (Cornwall, Wales and Scotland). More importantly, they brought new ways of managing land and, within a few centuries, had cleared forest from most of the best soils in England and brought them under the plough.[363] They soon appreciated that more was to be gained by working the fertile land than robbing each other.

The Anglo-Saxons also perceived that, instead of living in a vast continent where they had always been at risk from invasion because there were few natural boundaries, they were on an island with impoverished and uncooperative tribes to the west and north, and the sea to the east and south. This geographical isolation gave them a measure of security, and they visualised a more peaceful future. Even though there were long periods of war with Scandinavians, who had settled the north-eastern quarter of what was to become England, the Anglo-Saxons soon found that the Scandinavians had come to similar view of the future. Thus, it was during periods of relative stability after the victories and negotiations of King Athelstan (*ca.* 894 to 939 CE) that the Common Law of England evolved. That basis of law was founded on the needs of free peasant landholders who had no claims to nobility and owed service or dues to no one below the king. Thus the king became the symbol of law and order.

Each family farmed a *hide* of land (40–120 acres, depending on the landholder's abilities) and served in the *fyrd* or territorial army.[364] They paid *feorm* (food rent) to the king who, in turn, paid his bodyguard and other supporters.[365] Feorm, which gave us the word 'farm',[366] had most likely evolved from the 'protection money' farmers had been forced to pay, from the beginning of the Neolithic, to travelling bands of robbers who, if they were successful, later made themselves kings.

Private Anglo-Saxon households were grouped together into villages, and villages were combined into areas that were variously called *hundreds* or *wapentakes* (an Old English word denoting the shaking of weapons by which men confirmed decisions at meetings).[367] To maintain the evident prosperity of peace, when individual peasant landholders disagreed over a matter, they turned for an opinion about rights and wrongs to a third party whom both trusted, rather than settle the matter by fighting. If either of the disputants disagreed with the judgement, they took the case to a larger body of peers. Crimes were also dealt with by a public assembly or *folkmoot*.[368] Later the largest landholder or 'lord' took responsibility for organising decisions and collecting the fines, though he was answerable to the king's *ealdormen* (aldermen) and shire-reeves (sheriffs). Seventy years before the Norman Conquest, 12 leading thanes of the wapentake swore that they would neither protect the guilty nor accuse the innocent.[369] The modern jury system evolved from this. Today, lawyers have to convince 12 citizens of their case, relying on clarity of exposition, convincing evidence and the models already in the jurors' minds.[101]

Judges used the reasoning behind previous judgments to inform their decisions in cases that were similar. In this way, local decisions accumulated and became incorporated into a body of law. Many people in the course of many trials carefully examined such court decisions before they reached the statute books, and this imposed a strong selective process on them. (Darwin, again.) Roman or Civil Law is not subject to such refining because they are drafted and scrutinised by a few people only. The Anglo-Saxon landholder's main interest, and the subject matter of most disputes, was largely to do with personal freedom and security, property and contract rights.[370] Thus, Common Law evolved from grass roots upwards, and ordinary people felt secure in their title to land, their ownership of goods, and the future well-being of their children.[371] The common people had agreed the law and passed it to government

[101] The last of these was usually the most difficult to overcome, as it is in the case of this book's central thesis.

to engross. So deeply entrenched was this form of social organisation in Anglo-Saxon society, that even William the Conqueror knew that, to avoid lifelong resistance to his rule, he would have to swear to uphold the Common Law.[372] The Norman occupation was socially a dark age in England for two main reasons. First, England became an outpost of an empire that at one time stretched from the Mediterranean to Hadrian's Wall and, like the Romans before them, the Normans looked abroad to their focus of allegiance and repository of plunder and taxes. Second, the invaders brought with them preponderant Common Law. The end of Norman occupation, and their absorption into England, was sealed 800 years ago in the *Magna Carta*, whose substance is essentially Anglo-Saxon, and re-established the Common Law.

Later in the Middle Ages, the Roman Catholic Church, in the person of the Pope in Rome, issued edicts that affected the lives of people in distant lands. This was accepted by nations under Roman law but not in England where Common Law ruled. Apart from his marital problems, King Henry VIII perceived that the people would back him, if he made the established church local. That was an important consideration in his decision to found the Anglican branch of Protestantism with himself at its head. We see similar political power struggles today between Westminster and Brussels. People brought up under Common Law are suspicious of any power beyond their shores because they believe that loyalty to common well-being overrides all others, especially those who seek to abuse it for personal gain. Most people have confidence in the clumsy and inefficient living organism that is English Common Law and its courts. This is why *fatwas* issued in Mecca and attempts to set up alternative courts based on Sharia law elicit such outrage. And the offence is magnified by those whose dress defiantly proclaims such foreign allegiance and contempt for agreed local laws.

Of course, strong elements of Common Law remained in the unwritten English constitution; dictates of the king, who also commanded the army, persisted after times of war and were grafted onto the matrix of Common Law. They had more to do with matters that affected the nation as a whole, such as international trade and treaties. The English type of democratic institutions is sufficiently robust and flexible to allow the election of dictators in time of national emergency, such as World War II. The electorate were confident that, when the crisis had passed, the dictators would leave office peacefully. However, that is not a universal attitude. I was in Morocco during the election of a US president, and asked a local man what he thought of the process. It was incomprehensible to him that anyone, who commanded the most powerful army in the world, should voluntarily relinquish power. I did not try to explain that the fact of a peaceful change of president was a main reason for that army becoming so powerful.

By the 13th Century, the English Crown had delegated its power to three royal courts, all sitting at Westminster Hall: the Exchequer – revenues of the Crown; the Common Bench – suits between subject and subject; and King's (or Queen's) Bench – criminal cases or civil cases affecting the Crown.[373] Later a plethora of other courts appeared and they have evolved into the highly complex and cumbrously effective legal system that operates in England today.

England eventually bequeathed the Common Law base of her legal system to the USA, Canada, Australia, India, Malaysia and much of Africa, indeed all the present or former colonies, protectorates and dominions of the British Empire though a number have reverted to less democratic systems of government. The continental countries of Belgium, France, Germany, Portugal, Spain and The Netherlands, which had retained an administrative inheritance from the Roman Empire, gave a Common Law basis to their colonies in other parts of the world.

Only Common Law has always been intended to apply equally to everyone; indeed, equality under Roman or Common Law was unlikely to have been the original intention. Lawgivers often regarded themselves as being above their creations as illustrated by President Nixon's words: 'Well, when the

President does it, that means it's not illegal'.[102] This attitude creates an important difference between top-down Common Law and bottom-up Common Law. The difference is that the former is more open to takeover by a kleptocracy whose personal interests override those of common wealth. Roman law has led to centuries of mistrust and insecurity of property rights among poorer people – notably the *latifundia* still widespread in Spain; and this, in turn, undermines tenants' confidence in having secure futures.[103]

**********

Just as the quality of weapons is an important factor in deciding the outcomes of hand-to-hand battles, so debating battles are fought with words. Having the support of well-known and articulate people counts more than the same number of less impressive individuals. Masses can be swayed by good argument and strong personalities – and also, on occasions, by wealth. At the end of a Saxon verbal contest, results were judged by the shaking of weapons. Each contestant could see not only how many men were in favour of him but also what sort of weapons they carried. A dilemma emerges here: do debaters aim for numbers of followers or their quality? Obviously both, if they can get them, but usually it is a choice.

Today we try to do things more peaceably and equably in that each person has one vote. This fitted with the idea of each human being having a soul which was counted as equal in the eyes of a god; it certainly chimed with our innate notions of justice and fairness, and everyone being equal under the law. But there may be a more insidious force at work beneath the surface. Much of politics is by debate, and politicians' jobs are determined by voting. Because fewer people are well informed and intelligent (able to modify their behaviour to achieve a stated aim) than others less well endowed, many politicians

[102] Richard Nixon interviewed by David Frost, May 1977, reported in the *Guardian* on 7 September 2007.

[103] I once went with a landowner to collect rent from a goatherd who lived in great poverty in the cork oak forest of Andalucía. The rent was half of what the goatherd had earned that year. I did not repeat the experience.

perceive that it is easier to convince more people of a small distortion of argument or fact than few of a large one. If politicians seek power to further their own personal interests rather than the common good, they would be sensible to aim for masses rather than thinkers, and they can do this effectively by appealing to the most common emotional factor to be found in potential supporters.

An effective way of reducing the impact of well-informed and intelligent opponents is to arrange that each person has only one vote. The right to vote was historically confined to certain classes: women were long excluded and children still are. Now adult franchise is practically universal, though certain sects of Islam are seeking to reverse this. Votes are cast in geographical constituencies, which, in England, originally followed areas of judicial administration.

Two fundamental issues here are: should the differences between voters be recognised, and what is a constituency? First, voters. Throughout the history of life on Earth, evolution has operated on the basis of no two individuals being the same. During the last century or two, it has become more and more politically incorrect to state the obvious truth that we are different.

There is an urgent need to overcome Callicles's conundrum. Callicles argued in conversation with Socrates that, in a democracy, the majority are weak and make laws to control the strong. He did not say what he meant by 'weak' – bodily, socially, economically, intellectually? If the last of these is widespread, we have a problem. It is this: people who have read widely and thought deeply about civic duties and their contribution to society, feel resentment when they realise that their votes have exactly the same value as those who have never read nor scarcely thought about the issues, and whose main ambition, it seems, was seeking sensory gratification.

The problem can be mitigated, if not solved, by relating experience and intelligence to voting weight.[104] There is a way of defeating the proposition that the majority is always right, and it is by recognising the unassailable truth that we are not equal when we make political decisions.

Without more preamble, I propose the following system for England, with appropriate conversions for other countries: at age 16 every citizen is allotted 16 votes, and they gain one more each year until they reach 70, then they start losing them at the rate of 2 a year. English school pupils sit a public examination called General Certificate of Secondary Education (GCSE) at age 15–16. In each GCSE subject, candidates would gain 3 extra votes (to be used when they were 16) for an A grade, 2 votes for a B, and 1 for a C. Similarly, for the public examination sat at age 17–18, which is called 'Advanced Level' in England, the scale would be tripled: 9 for an A grade, 6 for a B and 3 for a C.[105] Those who went on to university would earn 60 votes for a

---

[104] This idea is not new. I first came across it in Nevil Shute's novel *In The Wet* (1953).

[105] Having spent much of my life teaching, I had intended to include a chapter on education but realised that there is not time enough for any suggestions I might make to take effect before the population tsunami was upon us. Even so, the eight years I spent examining both A and O levels for Oxford Local Examinations convince me that the grading system is deeply flawed. With what insufferable arrogance do examiners think that they can mark and award grades to a constant level, let alone set an examination that is uniformly difficult from year to year? The only worthwhile constant in an examination is the average ability of candidates in a large-entry subject.

An examination fails in its purpose if no candidate can answer any question correctly; it fails equally if all candidates get all questions right. The purpose of an examination is to *distinguish between* candidates so that end users of the results – selectors for further education or employment – can have some assessment of candidates' suitability. Accordingly, the only duty of an examination, however hard or easy, is to apply a constant standard to the achievements of each candidate in that particular examination. Candidates are then placed in rank order according to the marks they have received, and this position is shown on their certificates. Why should one who has come first out of 20,000 candidates not be recognised? Why should examiners take it upon themselves to decide that another candidate who came 8,347th should receive a C grade, while a third who came 8,348th should receive a D grade? By replacing grades with

first class degree, 40 for an upper second, 30 for a lower second and 20 for any other class. Maybe there would have to be some sort of scaling between universities to distinguish between those of recognised academic rigor and others less so. There would be a similar scale of votes for professional qualifications. Each £1,000 of income tax paid on earned income between general elections would gain more votes up to a maximum of 100. Maybe there are other categories that ought to be considered but, like the above, they should be accurately measurable and connect experience, thought and contribution to society with voting weight. Certainly the figures I give above are to be argued over.

The problem is how to get such measures into the electoral system. While people perceive voting as little more than a device for extracting the maximum personal and selfish advantage ('everyone votes in their own interest'), there is little likelihood of the idea of multiple voting gaining ground. Certainly it is unlikely that it would be received by a top-down government with anything but derision. I see no way of instituting it other than by starting contentious discussion and persuading voters that our real interest is in perpetuating, in generations that follow, what we are and what we have achieved. We all need to perceive that immortality is something we can do now while we are alive, rather than an automatic process that ends up in some mythical paradise or hell. There is a certain attraction in the notion that we would be remembered favourably by generations that follow. In short, we need to spread the idea of bottom-up government, as practised by countries that use Common Law. It is a race against time because population pressure is beginning to turn nasty. But we

---

actual positions (converted to percentage for end users weak in arithmetic), examinations would hand over to end users the responsibility for looking at candidates' other attributes. Examinations have become so contaminated with anti-elitism that grades are fast becoming political appointments. There was a glimmer of hope in the Olympic Games of 2012 that inspired a generation with elitism. After all, we have had 3.5 billion years of it in our evolution. See my letters to *The Independent* of 3 September 1992 and *The Daily Telegraph* of 13 September 2012.

have modern communications to spread ideas. I see no other way of starting these changes.

**********

Just as the concept of one person, one vote emerged from our historical past and was sanctified by religious ideas about a soul, so also has the linking of political constituencies with the land emerged from historical conditions of the past. We are no longer more or less evenly distributed over the land as we were in pre- and early industrial times: we are more concentrated into cities than ever before.

Responsibility for many local services has, in the past, been progressively taken over by central government, but there is, in 2015, a trend towards returning that responsibility to local authorities. Thus, local government should continue to be elected on a territorial basis. But issues such as the economy, law and order, foreign affairs, defence and measures to reduce population are national matters and should be the concern of politicians with different priorities. Modern communications are such that no longer do masses turn out to attend hustings, but prefer to watch proceedings on television. This too undermines the need for national politicians to be geographically bound. Individually, we probably know more about our national politicians than our local ones. Accordingly, I suggest that weighted voting should be introduced to elect members of a national parliament whose duty it is to deal with national interests.

At the same time as introducing weighted voting, the bases of constituencies for election to central parliament should be changed from territory to shared interest. Thus voters would have to choose what their individual interest-constituency would be: teacher, engineer, administrator, military, food industry, academic, retired folk, candlestick maker, and so on. Each constituency would have seats allocated to it according to the number of voters who have declared their choice on the electoral roll.

**********

Before going further, we need to look at institutions of state other than the legislature. In England, a police force grew out of the judiciary. The first policemen were the Bow Street runners who were sent to apprehend suspects wanted by Bow Street Magistrate's Court in London. Before that there had been an ad hoc system in which a magistrate read the Riot Act to quell civil disturbances. Failure of the crowd to comply with the Act, entitled the magistrate to call out the army. In other countries, offshoots of the armed forces became the police force directly.

A legislature, a judiciary, a police force, and the armed forces needed an infrastructure of administrators: a civil service. Together with an established church, these six bodies formed the great institutions of state. But all of them were staffed by people with the usual human frailties and ambitions. In an attempt at forestalling corruption or even usurpation, the monarch insisted that all individuals who served in those institutions swear allegiance to the institution of monarchy.

That such an oath is not taken in England today at every induction is a measure of misunderstanding how the role of English monarch has changed. In the past, absolute monarchs claimed to rule by divine right, and exacted taxes in exchange for defending their people against external threats (who were often the monarch's own personal rivals); in addition, they took on the role of avenger:[106] redressing wrongs and defending the mass of people against oppression. Still today, oppression comes from three sources: invasion by other nationals, criminal acts of fellow citizens and unjust interference by the monarch's own agents – the legislature, the police, the armed forces or the civil service. Of course Queen Elizabeth II does not take up arms and go into the streets herself, but she represents the law. That is why legal cases involving the state of Great Britain and Northern Ireland are always called *plaintiff/defendant v. Regina*. In other words, the modern constitutional monarch stands between law-abiding citizens and oppression. And it is also the

[106] The Bible records many exhortations to leave Vengeance to the Lord. See, for example, Romans 12.19, Leviticus 19.18, Deuteronomy 32.35 and a host of others.

reason why, in a large part, constitutional monarchs have survived into the modern age.

These are the reasons why all serving members of the six great institutions of state listed above should swear an oath of personal allegiance to the Crown. Not to the individual who currently wears it, but to the institution embodied in the current monarch and all her successors. The Crown cannot be an inanimate symbol; it has to be a living person who, with advice, can take decisions, and it matters not at all whether it is a woman about to be ninety or a man in his sixties or thirties. In effect, swearing such allegiance is to uphold the Common Law and to guard the individual freedom of every citizen.

The role of constitutional monarchs as heads of state also includes final authority within the state, and they hold it so long as they never use it capriciously. That power has been used in modern times when the monarch, in 1975 and in the persons of Queen Elizabeth II and her Governor General, dismissed the prime minister of Australia and ordered a general election, which the opposition won by a landslide. This final authority and power can only be held by one person who must be above the institutions of state, must never comment on political or judicial matters, and must be separate from the common people. That person cannot, therefore, be elected, nor can they be appointed. The only alternative is to breed them.

And this brings us back to why men struggled to be kings. But thenes have changed all this; most of us have evolved away from simply multiplying DNA and come to regard our possessions as important as our genes. For a few of us, idenes are even more important. And this is another way of saying that our individual behaviour is more important than our race, our genealogy or our possessions. It does not really matter if a monarch's parentage may be suspect,[107] so long as they appear to have been bred suitably, they accept the job and behave appropriately.

---

[107] Wilson A. (2003: pp. 24 to 25) casts doubt on Queen Victoria's biological pedigree.

**********

It has been said that democracy is the worst form of government except all those other forms that have been tried from time to time.[374] But democracy depends on the electorate accepting the will of the majority. However, more and more people are becoming reluctant to do this, especially when margins of victory are narrow. This resentment may have some of its roots in the cult of winning in sport: some people see losing as a situation to be avoided at all costs, and to be excused or fought against if shades of doubt about manhood (and it is nearly always men who respond in this way) are to be cast off. Watch any professional association football match to see heroes grovelling in apparent agony while trying to 'win a penalty'. At the school I attended in the 1950s, there was a home athletics match against another, similar, school. In the high jump event, the favourite to win – a visitor – waived aside the early rounds when the bar was set, he considered, too low for his dignity. Eventually he deigned to try, wearing his tracksuit, and knocked the bar down. A spectator pupil laughed. The school prefects, who administered school rules in those far-off days, sent for that pupil and caned him for displaying such ill manners to a visiting team. Outrageous, by today's standards, when the arrogant high jump oaf deserved to be hooted off the field, but the story still carries a message. Less than 30 years later, I watched the track events at an athletics match in another school, and noticed that the majority of competitors who came last in an event, limped off the track.

Today's media perceive winning as being a greater honour than taking part. There is surely more respect due to participants in Olympic Games than that they won a medal of whatever metal and by whatever fraction of a second. It is no slur on manhood or disgrace to be beaten by a rival who is clearly better than oneself, even if 'better' in this case means 'having more votes', as in my proposed voting system. Losing with dignity is a sign of individual maturity and magnanimity. Not doing so is merely a pathetic display of immature emotion. More sinister than that, sportspersons that behave like this set role model examples to young people that act like dry rot in the foundations of democracy, which depends on accepting the will of others. Losing is a vital function of a democracy because, being composed of fallible human beings, every executive needs a vigorous, sceptical and inquisitive opposition.

**********

But politicking alone will not make the practicalities of population reduction less daunting. In preparing ourselves to think about the practical steps that our refusal to build a new London every five weeks would entail, two more major issues need attention: reputation and punishment. First, let us consider reputation.

We evolved in small tribal groups that settled down in villages when we invented agriculture. Everyone in the hunter-gatherer group or the farming village knew one another and had done so from birth. Each individual's biography was attached to their distinctive face. It was easy to know who was to be trusted and who not. Gossip was the stuff of village life, scandal and rumour were rife. Social conformity and observance of local rituals were essential. Language evolved, at least partly, to manipulate collective memories of our biographies.[375]

Rumour-mongers were soon identified and that propensity was added to their own records in the living history book that is village memory. The power of such a system in moderating behaviour was enormous. Only children were forgiven their sins, and even then there were exceptions in severe cases. I once went to a remote Shetland island and met a senior and respected inhabitant, but it was soon disclosed to me by others that, as a child, he and his brother had set fire to the peat one summer. Peat was the islanders' winter fuel for generations to come, and they nearly lost it: the fire took weeks to put out, and at enormous cost in labour. Had an adult committed such a crime, I am sure that he would have been forced to leave the island. This misdemeanour was in the islanders' oral history.

Our migration into cities vastly increased opportunities for selfishness, antisocial behaviour and crime: our faces were unknown; we were anonymous. The chances of anyone recognising us, let alone our being caught, were small enough to outweigh the benefits of criminal actions. Now, in the last few years, we have evolved the capacity to make the world an electronic village: details of criminals can be called up in seconds from any part of the world. As a resident of Spain, I had to carry an identity document (ID) that showed my face, fingerprint, signature and address. The hysteria surrounding the introduction of IDs in the UK is remarkable. What do objectors want to do that would be inhibited by their ID? Freedom or privacy is not enough, and secrecy suspicious.

At the same time, and I think as a reaction to urban anonymity, there has been a vast and voluntary increase in social networks, such as *Facebook*, where people announce their identity and all sorts of, sometimes embarrassing, personal details. The important point about these networks is that those who join them decide what information shall be displayed and also who shall see it. Social networks include email, chat rooms, twittering, blogging and more general websites supplying opportunities for research into the lives of other people. The result is that almost instantaneous communications have created a virtual global village, similar in some respects to the Stone Age communities we evolved in.

Communal censorship, individual surveillance and malicious subversion have been practised since early hominid days right up to modern China.[108] It is hard for citizens of open and relatively free societies to understand how all-pervasive is the Chinese government's control of what its citizens are allowed to do; central records are kept and human freedoms seriously restricted – by western standards.

Suppression of *any* idenes reflects the gene- and thene-based mentality of those in authority: they regard the spreading of their own individual genes and the accumulation of wealth as more important than other people's ideas. All that will remain of such regimes, long after the ephemeral pleasure of their earthly wealth has disappeared, are the individual names of the perpetrators, recorded unenviably in history and public memory.

[108] Link (2011: p. 52): it is worth reading the whole of this article.

There seems to be a direct relationship between the degree to which a people worship their ancestors and the trouble their leaders take to make such bad ones.

What can really overwhelm dictatorial regimes is the sheer volume of text messages, tweets, blogs, and so on. These signals not only unite people of common purpose but also pass their individual and collective ideas to the world outside the repressive regimes in which they languish. The crucial point here is that it is the substance of the ideas so presented that attracts interest rather than the senders' desire to propagate them. Truly, Dawkins's suggestion that ideas can take on lives of their own is immensely powerful: idenes multiply, diversify and are selected, so evolving like living organisms. The essential criterion is that idenes are acceptable to a freedom-hungry people. When they are, they spread with disconcerting rapidity; disconcerting, that is, to the oppressive government.

**********

So effective are the ways of gathering, storing and transmitting information electronically that it would not be difficult to create an official and comprehensive database of information like that retained in the communal memory of ancestral villages. If official, compulsory, comprehensive and freely available websites with a page for every individual were to be set up, the social effects would be vast. How would we decide what should be included and what left out? Could we edit our own pages? What would be the sanctions against those who tried to falsify their own page, or tampered with those of others, if only by bearing false witness against their neighbour? Gossips and snoops would have a lean time because all they had to offer was public knowledge anyway. Would there be a risk that some people would object to others' behaviour, even though it were not illegal, and act upon it? These questions would be best answered by courts of Common Law. Childhood misdemeanours would be forgiven and erased at age 18, or the age of voting franchise; thereafter, criminal record would be public and permanent.

Punishment is concerned with the future, not with the past: it is for deterrence not revenge. Knowledge that conviction of a crime would mark an individual for life, and that there was no question of being able to pay off any debt to society other than by suicide, would be a serious discouragement to criminal activity. There is a risk that such a system would place people in the position where they felt that they had nothing to lose, and so act desperately; it is a risk worth taking, especially if the felons want to surrender their lives. Also recorded on each person's individual web page would be all the positive things that they had done, including reparations they had made for crimes or misdemeanours.

As things are at present (2015), lenient sentences and the probability of criminals remaining undetected causes frustration and discontent in victims of crime, and also in uninvolved citizens.[376] This can be to such an extent that they lose confidence in the law and in public institutions, feel contempt for both and are inclined to take enforcement into their own hands. Such social danger will surely become magnified as population pressures bear down on orderly behaviour, and also on the legal systems that presently seek to support freedom.

**********

The DNA-driven 'golden rule' of our doing to another person only what we would like them to do to us, runs the risk of influencing our judgement when we have to make decisions that appear detrimental to another human being.[377] We are reluctant to hand down severe penalties when we imagine what our fellow human being might feel about them. We need to balance our rational thoughts about what we can see threatens us against these genetically based feelings. With increasing overcrowding, severity of punishments should rise steeply for second and third offences. For example, if someone were found guilty of hiding money in excess of the £10 million they were allowed, and this was their first offence, everything they owned would be confiscated by the treasury and they would be given a small allowance to live on. A second offence would attract imprisonment.

Criminals have taken positive decisions to live outside the laws of the land they live in. In damaging the freedom of others, they have forfeited their own. Many of them simply cannot imagine what life would be like in a lawless society. It would help them to understand the civilised basis of living together under the rule of law if they were exposed to a lawless situation. With no sense of revenge but only the intention of altering future behaviour, I suggest that reoffenders be sent to a prison built of indestructible materials, and with guards but not warders. The prisoners are simply told to make their own rules about accommodation, sanitation and the distribution of food that is supplied daily. A week of such a regime would be salutary. They would quickly learn that freedom involves responsibilities, just as rights involve duties. First offenders would have secure apartments where they could see into the lawless enclosure and watch what went on in it.

**********

And that brings us to the most difficult and contentious part of this book. It is one thing to detach suicide from disgrace, to withdraw medical aid from certain age groups, and to terminate pregnancies, but it is quite another to consider deliberately taking someone else's life, but consider it we must. There are times when taking the life of a fellow human being is seen to be acceptable and sanctioned in law. Killing healthy young men in war is part of the deal, but there are often enquiries when unarmed citizens die, especially women and children. Police are expected to kill armed people who appear to be trying to kill them or others. So deliberately taking human life *can* be done under the law. With rapidly changing population and environmental circumstances, we need to discuss expanding the categories of people it would be lawful to execute. Maybe it would make courts' decisions less painful if they were required to decide only whether criminals were still entitled to call themselves members of the ethospecies *Homo sapiens*, as I discussed in the previous chapter. In practice, executing a few individuals would have a negligible effect on population. Its intention would be to alter behaviour by fear of such a possibility.

Perhaps the greatest threat to human security and freedom will be caused by the breakdown of the concept of property. As in Neolithic times, people will be forced to choose between dying of starvation and taking food when they see an opportunity. When supplies fail, towns will quickly run out of food, and the surrounding countryside will be the nearest place where there is some left, if only as growing crops. The law would have to be severely enforced to avoid the complete breakdown of civil order.

We have not reached that stage – yet. Meanwhile, there is no legal external compulsion to commit a crime; thus all criminals are volunteers. If a person in a developed country is so desperately poor that they think they will starve if they do not steal, they can swallow their pride and ask for assistance from family, friends or their local municipality. Yet still many people feel compelled from within to commit crime. Crime is probably no more common per head of population today than it was when there were fewer people, but it is certainly more common per unit area of land, therefore it is less tolerable. Managing prison populations will become even less affordable as the present bloated and environmentally destructive global economy begins to decline, as it surely must.

There are many factors that encourage crime, and one of the more pervasive is an unexpected consequence of that same branch of lekking that we can label 'look-at-me-caring'. In many developed countries, politicians have sought and gained election on the platform of being seen to care for the voters. How often do you hear the electioneering mantra 'hospitals, school, jobs, social benefit and pensions'? With vast modern welfare protection, and increasing restrictions on what the citizen is allowed to do – in the interests of health and safety, of course – more and more responsibility is being taken on by the state.[109] With diminished responsibility and freedom, the spirited individual revolts and looks for fresh boundaries to press against; where better than crime? It is exciting, challenging and,

[109] Aristotle warned against this.

above all, profitable; not to mention that, under the present legal and political climate, the probability that one will be caught is vanishingly small, and if caught, convicted, and if convicted, receive a sentence that deters.

Another factor that fails to discourage crime and spins off from the concept of human rights is criminals' knowledge that they can gain much delay, if not protection from it. Clearly no one is guilty until proved so, but a previous conviction should cause a criminal to forfeit their human rights. A citizen coming to court and claiming protection under these acts must have their hands clean.[110] Apart from rethinking how we should define both the nation state and our species, three major shifts in the relationship between state and citizen are urgently needed. The first is that the state returns to citizens the responsibilities that make them individuals: we are who we are, genes, upbringing, accidents, diseases, achievements, triumphs, warts and all. No longer can we individually blame our genes or upbringing for the way we behave. We have a duty of responsibility for, not to, ourselves first.

With these ideas in mind, I propose that a United Nations Declaration of Human Duties replace their Declaration of Human Rights.[111]

Everyone has the duty to:

1. accept responsibility for themselves as they are genetically and behaviourally, regardless of past history and present circumstances;
2. care for themselves in such a way as to maintain their health: smoking, drug-taking, alcohol abuse and obesity being considered misdemeanours that debar offenders from state-financed medical treatment;
3. work so that they can provide a standard of living adequate for their well-being and that of their family;

---

[110] Extremists of known villainy seek shelter from laws that protect human rights when their highest priority is to repeal those same Acts.

[111] Aristotle expressed a similar code 2,500 years ago (Mill, 2006: p. 97).

4. learn the dominant language, the history, laws and customs of the land in which they declare their residence; and learn and understand the history of the human species from the beginning of time;
5. obey the laws of that land;
6. pay taxes in that land to provide for the education, medical care and security of everyone in their community; and social services, unemployment benefit and pensions of those incapacitated through no fault of their own;
7. take active interest in elections and vote at them;
8. care for the environment and order their life so that they have minimal impact on the natural part of the world;
9. surrender their life when they can no longer contribute to the well-being of others.

**********

All this theorising is fine, or not, in itself but we still need to face up to practical considerations. No politician seeking re-election, or civil servant wishing to keep their job, would accept suggestions as outrageous as mine today. The only way that change will come is from the bottom-up, like Common Law. I see no way in which these logical measures can become effective unless they are spread by ordinary people thinking and talking about them, refining the arguments, correcting my errors and spreading the word.

'Spreading the word' – not like some gospel that is to be believed, but as an argument to be criticised. Only then could the ideas work their way upwards through layers of civil administration until they reach bodies that are capable of taking decisions and making laws on behalf of nations or continents. Once they have reached that level, the ideas suggested earlier should be processed through a reorganised UN in which veto by one is abolished and Callicles's conundrum is solved. Perhaps a system of multiple votes for the representatives of each country would work. An alternative approach would be through the most powerful economies like the G8, expanding to the G20, and so on. Once agreement is reached with these narrower circles of nations, assistance, sanctions and intervention become more feasible on a global scale.

**********

At long last we come to insisting on a final decision about whether we should or should not try to build, equip and staff the equivalent of a new London every five weeks, and *then* try to improve the lot of those who do not have an adequate standard of living. The decision is a matter of individual conscience. It is a choice between obeying our DNA-driven feelings which are wholly concerned with the present, or heeding our rational thoughts which can guide us into the not-too-distant future with measurable certainty.

Because it is unsustainable, our human population will certainly decline anyway. We need only decide whether we wish to cooperate with each other to make that decline happen under the rule of law, however savage those laws may have to be, or whether it will happen in a holocaust of lawlessness.

If we choose to do nothing, the first casualty will be the concept of property. In 2015, we are already seeing it in the mass migration of people leaving Asia and Africa to enter Europe. The land of Europe is the property of those who have lawful title to it. Masses forcibly entering to share in that ownership are breaking down more than fences, they are breaking down the civil contract by which we live peacefully together. It is a contract of cooperation and agreement and acceptance of things that may displease some of us individually. The scenes of migrating people we have witnessed on television during 2015 are nothing compared with what will certainly happen when import and export fail in the ecosystems that are each city: empty supermarkets and drains backing up. To what lengths will you who read these lines go to find food for yourself and your family? Will you watch them starve or will you steal for them? Or will you press for laws that we must all obey?

There is (in 2015) plenty of food in the world, and the price of oil is low. We still have time to share them both without bloodshed, but that time is now diminishing so quickly as to be practically past. Another dimension of our choice is whether we accept a little discomfort and emotional distress now, or thrust vastly more onto the next generation. This is a deeply flawed alternative because, as I have shown, we *are* the next generation.

# Epilogue

## IN THE LONG RUN

Understanding is an essentially human activity. Information has always existed within matter whatever form it takes; it is also there in the way that matter behaves. We convert information into knowledge by describing it in our symbols. One set of symbols is words, another is mathematics, and both of them can be expressed in a variety of media. Once information has been converted into human symbols, we can begin to understand it. It seems such a pity that all our knowledge and understanding will be lost at the end of the Universe, if not before. Maybe not; let us see.

'*In the long run* we are all dead.'[112] Maynard Keynes was right, wrong, right, wrong, right, and perhaps finally wrong. Ah me! That slippery concept 'the truth' is wriggling again. Like J. S. Mill, **Keynes was right** in that the physical form of our present bodies dies and is buried or cremated and the material components scattered. But we now know that the really important part of us is not the atoms of which we are made and which are continually changing throughout our lives, but the information which orders how those atoms shall be arranged and how they will behave. Even though that information is fragmented as it passes on to our children, it is not destroyed; anyway, the vast bulk of it is carried by all other human beings. From this point of view, **Keynes was wrong**.

Our Sun is in middle age and still has about 3 billion years of life before it begins to die. When its hydrogen fuel starts to run low, its temperature will rise, oceans on Earth will boil away and mountains melt and run into the valleys. Eventually the Sun will cool and expand to become a Red Giant whose bloated dying body will engulf the nearby planets, including Earth. Then it will collapse and its temperature rise until all the atoms,

---

[112] Keynes (1923: p. 80), original italics.

including ours, will be crushed into proton and neutron plasma. No structure will be left and no information will survive: then we shall be truly dead. **Keynes was right**.

But he did not foresee the possibility of a spaceship. In 1977, deep-space probes *Voyager 1* and *Voyager 2* left the Earth and, having explored the outer planets, they continued beyond them. In July 2015 they were about 20 billion kilometres (km) from Earth, and they have left the solar system.[378] Travelling at 61,400 km per hour, one of them could reach the nearest star in about 74,000 years,[113] though neither is aimed in that direction. They contain gold-plated audiovisual disks containing some scientific information about the Earth, and some other trivial impressions. The chance of either *Voyager* meeting another form of intelligence that could decode the disk is just measurable, statistically; so **Keynes was wrong**, and he knew about probability.[379] In fact, the chance of the *Voyager* disks being read and understood is as near zero as makes no difference to ordinary mortals.

The universe we know is 13.7 billion years old, and its future stretches ahead while it expands and stars are born and die. It began at a singularity before which the laws of physics seemed not to have applied, then they came into being. Those laws suggest that we live in a universe of increasing disorder: matter steadily converting into energy, and that energy degrading into heat. Eventually our universe will wind down until all matter is mere heat: whatever is left of us will suffer 'heat death', nothing will remain, not even information. So **Keynes was right**.

Let us, just for these last few lines, release our imaginations from the rigor of reality and pretend.

> I am aboard HMS *Beagle*; the date is 21 December 1832.
>
> 'The evening was calm and bright and we enjoyed a fine view of the surrounding isles. Cape Horn, however, demanded his tribute, and before night

[113] Going back 74,000 years, *Homo sapiens* had not yet left Africa.

> sent us a gale of wind directly into our teeth. We stood out to sea, and on the second day again made the land, when we saw on our weather-bow this notorious promontory in its proper form – veiled in mist, and its dim outline surrounded by a storm of wind and water. Great black clouds were rolling across the heavens, and squalls of rain, with hail, swept by us with such extreme violence that the Captain determined to run into Wigwam Cove.'[380]

'Permission to speak, Mr Darwin, sir,' said I knuckling my brow.

'Yes, Jacoby, what is it?'

'Begging your pardon, sir, but I'm more than a little worried about our position. Cape Horn is somewhere away on our starboard in the mist and darkness, and I fear that, after a full year at sea, there is such disagreement between the twenty-two of Mr Harrison's chronometers[381] we have on-board, that we cannot know where we are to the nearest half mile. Do you think that I should approach Captain FitzRoy?'

'As you will understand, the Captain is very busy at this moment and will surely not welcome interruption.'

'Begging your pardon again Mr Darwin, sir, will you tell him that I know where we are to within ten yards?'

'Don't be ridiculous, you are perfectly mad to make such a claim.'

'No, sir, really; I have this little flat box with a screen on it, and you can see that it is marking out a trace of our passage. If I touch this symbol, the scale increases and you can see that we are four hundred and thirty-two yards, plus or minus five each way, from the foot of the Cape. If I decrease the scale of the map like this, you can see much of the south-eastern Pacific Ocean. You will also see that a depression is heading this way and deepening. A plan of the isobars will make it easier ... like this. The wind speed will rise as the isobars compress and veer to onshore in about four hours' time. If you still have doubts, sir,

I can touch these numbers in a certain order and you can speak to the duty officer at Greenwich, and he will confirm the exact time there, so we can check our chronometers.'

After my conversation with him, Charles Darwin would have been in a predicament: he had to decide whether I was insane, and so incur FitzRoy's wrath, or take a chance that I might have been right. I had presented him with certain pieces of information but he had no background knowledge into which he could fit them, no model with which he could compare these new impressions. Looking back over the 162 years that intervened between Darwin's visit to Drake Passage and my real one, the steps that led to managing radio waves, rockets and satellites are entirely logical to us. To Darwin, the gap was too big; yet his mind was structurally and physiologically no different from ours, so capable of all that ours can do.

I wonder if we in 2015 are not in the same predicament, with respect to other fields of understanding. Now suppose someone from 162 years hence came into the room today and started talking about 'Five'.[114]

'Five'? With what arrogance do we think that our relatively feeble sense organs, our paltry little machines and really rather narrow ways of thinking, which we try to express in our archaic, clumsy and emotional languages, have perceived or found evidence for the existence of all the basic dimensions of being? We have a working knowledge of space and time, mass and energy, plus whatever other hypothetical dimensions mathematicians invoke to support their equations, but we have no evidence of any others. Just because we have never come across a fifth dimension of being, gives no hint that it does not exist. Indeed there are plenty of events that we do not understand, and a lazy mental attitude is to dismiss them as supernatural, which is merely deferring an explanation.

Here is a list of some things we label as supernatural: calling dogs from great distances,[382] clairvoyance, dowsing, extra

---

[114] A neutral name with no implications of energy.

sensory perception, faith healing, ghosts, premonitions, prescience, radionics, religion and doubtless you can think of others. Of course, many of them can be dismissed as ineffective, faked or imaginary, or explainable in terms of the known dimensions, but I defy anyone to *prove* that there is not a dimension beyond what we know about. If there is, and that dimension is independent of the universe we understand, and we could communicate our knowledge through it, perhaps **Keynes was wrong** after all.

In spite of Nietzsche's and Scruton's pessimism,[383] the human condition needs that last mosquito on which Pandora slammed the lid of her jar: it was called 'Hope'. We have a chance to explore it, if we prepare ourselves for the population tsunami and survive it in tolerable order. After that, there is plenty of time, because 162 years is not long, and we have at least 3 billion years before the Sun dies. And in exploring the idea of hope, I would rather consult a physicist than a priest, because the physicist has doubt.

# ENDNOTES

[1] The Universal Declaration of Human Rights. Available from: http://www.un.org/en/documents/udhr/
[2] See Polunin (1998: p. 139, footnote)
[3] Skidelsky (2003: ch. 19)
[4] Wikipedia article on *World population*. Available from: http://en.wikipedia.org/wiki/World_population#Antiquity_and_Middle_Ages
[5] United Nations 2015 Revision of World Population Prospects. Available from: http://esa.un.org/wpp/unpp/p2k0data.asp
[6] Ehrlich (1971) and Malthus (1960)
[7] Cohen (1995) and Ehrlich (1971: p. 108)
[8] *New York Review of Books*, 10 February 2011, p. 26
[9] Scruton (2005)
[10] Tinbergen (1951)
[11] O'Hear (1999: p. 163)
[12] Cardinal Newman quoted by Wilson E. (1984: p. 267)
[13] Wilson E. (1984: p. 34); see especially Reilly (2010)
[14] Friedrich Nietzsche quote. Available from: http://www.notable-quotes.com/c/conviction_quotes.html
[15] Harari (2011)
[16] Goodall (1990)
[17] Dunbar (1996)
[18] Levitin (2015)
[19] Graves (1955) and Murray (1994)
[20] Frazer (1924)
[21] Popper (2002)
[22] Google Images. Available from: http://images.google.co.uk/imgres?imgurl
[23] Abercrombie (1960: p. 28)
[24] Berne (1974: ch. 2)
[25] See especially Berne (1974) and Harris (1970)
[26] Winston Churchill. *Hansard*, 11 November 1947
[27] Hamilton (2001: pp. xxxv–xxxvi)
[28] de Montaigne (1995: p. 38)
[29] Eliot (1999: ch. 2)
[30] O'Hear (1999: p. 98)
[31] Leakey & Lewin (1995: p. 310)
[32] Wilson EO (1998: p. 105)
[33] Gould (2000: p. 61), Wilson E (1984: pp. 12, 58) and Wilson E (1998: pp. 32, 40, 65, 229)
[34] Hawking (2001: p. 78)
[35] Maynard Smith & Szathmáry (1999: p. 31)
[36] Reilly (2010)
[37] Pross (2012)
[38] Gould (1977: p. 40)

[39] Browne (2002: p. 420)
[40] Blunt (1971: p. 179)
[41] Browne (1995: p. 210)
[42] Darwin (1902: p. 42)
[43] Barrett & Palmer (2009)
[44] Mittermeier *et al.* (eds.) (2013: p. 14)
[45] Nichols (1962: p. 91)
[46] Darwin (1860: p. 490)
[47] Mithen (1996: pp. 227, 231)
[48] Byrne & Whiten (eds.) (1988: p. 4), quoted by Mithen (1996: p. 89)
[49] Allman (1994: ch. 5)
[50] Dunbar (1996: pp. 69 *et seq.*)
[51] Allman (1994: pp. 124 *et seq.*)
[52] Goodall (1990: pp. 4, 15) and Mithen (1996: p. 80)
[53] Diamond (1991: pp. 127 *et seq.*) and Dunbar (1996: p. 48)
[54] Mithen (1996: pp. 65 *et seq.*)
[55] Calder (1972: p. 53) and Leakey & Lewin (1992: p. 84)
[56] Leakey & Lewin (1992: p. 72), Mithen (1996: pp. 227, 231) and Romer (1933: p. 614),
[57] de Waal (1996: p. 30) and Goodall (1990: pp. 83 *et seq.*)
[58] de Waal (1996: pp. 136 to 146), Goodall (1990: p. 177), Leakey & Lewin (1992: p. 185), Ridley (1996: p. 109) and Sober & Wilson (1998: p. 142)
[59] de Waal (1996: p. 151)
[60] Goodall (1990: pp. 36 *et seq.*, 45 *et seq.*)
[61] de Waal (1998: p. 128)
[62] Diamond (2000: pp. 106 *et seq.*)
[63] Goodall (1990: p. 40) and de Waal (1996: p. 93)
[64] de Waal (1996: p. 44)
[65] Allman (1994: p. 61) and de Waal (1996: p. 101)
[66] Birkhead (2000: p. 149) and de Waal (1996: pp. 76 *et seq.*)
[67] Dunbar (1996: p. 91), Leakey & Lewin (1992: p. 302), Mithen (1996: pp. 166 *et seq.*, and p. 217), O'Hear (1999: pp. 123 *et seq.*), Ridley (1996: p. 142), Sober & Wilson E (1998: pp. 252 *et seq.*), Stamp Dawkins (1993: pp. 4 *et seq.*) and Wilson E (1998: p. 120)
[68] Dunbar (1996: p. 91), Goodall (1990: p. 18) and Leakey & Lewin (1992: p. 298)
[69] Mithen (1996: p. 217)
[70] Mithen (1996: p. 85)
[71] Diamond (1991: p. 29) and Leakey & Lewin (1992, see index)
[72] Dunbar (1996: p. 69 *et seq.*)
[73] de Waal (1998: p. 172)
[74] Mithen (1996: p. 169)
[75] Dunbar (1996: p. 107)
[76] Dunbar (1996: pp. 17, 22, 107)

[77] Mithen (1996, see index)
[78] Eliot (1999: p. 34)
[79] Diamond (1991: p. 33) and Diamond (2000: pp. 124 *et seq.*)
[80] Ridley (1996: pp. 49, 100)
[81] Dunbar (1996: p. 64). See also 'Does food sharing in vampire bats demonstrate reciprocity?' by Carter & Wilkinson (2013). Available from: http://www.ncbi.nlm.nih.gov/pmc/articles/PMC3913674/
[82] Allman (1994: p. 98) and Dunbar (2010: p. 24)
[83] Dunbar (1996: p. 62)
[84] Wikipedia article on *Facebook*. Available from: http://en.wikipedia.org/wiki/Facebook
[85] 6 new facts about Facebook. Available from: http://www.pewresearch.org/fact-tank/2014/02/03/6-new-facts-about-facebook/
[86] Leakey & Lewin (1992: p. 160)
[87] Diamond (1991: p. 53)
[88] Diamond (2000: p. 104 *et seq.*)
[89] Diamond (1991: p. 67)
[90] 'Scientists debunk myth of one in 10 children born to 'wrong' father.' The Telegraph, 11 February 2009. Available from: http://www.telegraph.co.uk/science/science-news/4586120/Scientists-debunk-myth-of-one-in-10-children-born-to-wrong-father.html
[91] Diamond (1991: p. 31) and Leakey (1992: p. 47)
[92] Leakey & Lewin (1992: p. 161)
[93] Mithen (1996: p. 138)
[94] Mithen (1996: pp. 132 *et seq.*)
[95] Leakey & Lewin (1992: p. 265) and Maynard Smith & Szathmáry (1999: p. 151)
[96] Diamond (1991: p. 126)
[97] Allman (1994: p. 178)
[98] Pinker (1994: p. 365)
[99] Leakey & Lewin (1992: p. 254), Maynard Smith & Szathmáry (1999: p. 150), Mithen (1996: p. 159) and Pinker (1994: p. 384)
[100] Pinker (1994: p. 40) and Ridley (1993: pp. 310 *et seq.*)
[101] Eliot (1999: pp. 27 *et seq.*)
[102] Pinker (1994: p. 406)
[103] Maynard Smith & Szathmáry (1999: p. 166)
[104] Dunbar (1996: pp. 79, 148 *et seq.*) and Ridley (1993: p. 227)
[105] Dunbar (1996: p. 123) and Mithen (1996: p. 161)
[106] Dunbar (1996: p. 176)
[107] Allman (1994: p. 108)
[108] Allman (1994: p. 37) and Dunbar (1996: pp. 168 *et seq.*)
[109] Cartwright (2000: p. 310), Dunbar (1996: p. 146) and Ridley (1996: p. 326)

[110] Dunbar (1996: p. 146)
[111] Eibl-Eibesfeldt (1996: p. 77)
[112] Dunbar (1996: p. 61)
[113] Pinker (1994: p. 286)
[114] Pinker (1994: p. 286) and Encyclopaedia Britannica (1991: v. 14, p. 717)
[115] Flannery (1994: p. 301)
[116] Dunbar (1996: p. 128) and Mithen (1996: p. 220)
[117] Dunbar (1995: p. 48)
[118] Diamond (1991: p. 37) and Mithen (1996: pp. 147, 213)
[119] Chicago, The Guardian, 12 February 2009
[120] Diamond (1991: p. 45)
[121] Marean (2010: p. 60)
[122] Marean (2010: p. 55)
[123] Mithen (1996: pp. 129 *et seq.*)
[124] Blackmore (1999: p. 315)
[125] Dawkins (1989: p. 201)
[126] Diamond cited by Donovan (1989: p. 247), Leakey & Lewin (1995: p. 175) and personal observation Galápagos 1996, 2004
[127] Dunbar (1996: p. 158) and Mithen (1996: p. 203)
[128] Diamond (1991: pp. 305 *et seq.*) and Donovan (1989)
[129] Donovan (1989), Flannery (1994) and Leaky & Lewin (1995)
[130] Allman (1994: p. 207) and Andrews (1991: p. 30)
[131] Flannery (1994: p. 57)
[132] Leakey & Lewin (1995: p. 174)
[133] Goodall (1990: p. 175)
[134] Livingstone Smith (2007)
[135] Goodall (1990: p. 85)
[136] Andrews (1991: p. 46) and Clutton-Brock (1999: p. 19)
[137] de Waal (1996: p. 137) and Eibl-Eibesfeldt (1996: p. 188)
[138] Ridley (1996: p. 116)
[139] Diamond (1991: p. 168)
[140] Diamond (1991: p. 264)
[141] Diamond (1998: pp. 268 *et seq.*) and Eibl-Eibesfeldt (1996: p. 225)
[142] Diamond (1998: pp. 99, 153)
[143] Terborgh (2007: p. 35)
[144] Diamond (1998: pp. 242 *et seq.*) and Pinker (1994: p. 272)
[145] Spindler (1994: p. 252)
[146] See also Pinker (1997: p. 509 *et seq.*)
[147] Flannery (1994: p. 150)
[148] Flannery (1994: p. 182)
[149] Flannery (1994: p. 225)
[150] Encyclopaedia Britannica (1991: v. 29, p. 433)
[151] Leakey & Lewin (1995: p. 174)
[152] Goodall (1971, endpapers)

[153] Cartwright (2000: p. 296)
[154] Eibl-Eibesfeldt (1996)
[155] Wallace (1869: p. 381)
[156] Bible: Matthew 9.9 and Mark 2.14
[157] Encyclopaedia Britannica (1991:v. 29, p.1029) and Pinker (1994: p. 200)
[158] Diamond (1998: p. 235)
[159] Keegan (1994: p. 122)
[160] Dunbar (1996: p. 62)
[161] Kennedy L (1999) and Mithen (1996, p. 200). See also Dennett (2006) for a fuller treatment.
[162] Lewis-Williams (2004)
[163] Lewis-Williams (2004: p. 176) and Matthiesen (1965: p. 139)
[164] Wilson EO (1998: p. 27)
[165] Ishaq (1955: p. 104)
[166] Bible: Luke 12.1
[167] Kennedy L (1999: p. 19)
[168] Frazer (1924: p. 9); Taylor (2012: e.g. p. 52)
[169] Stenton (1947: p. 363)
[170] Dawkins (2006: p. 344)
[171] Twenty-ninth verse of the ninth Sura of the Quran from Ali (1934: p. 447)
[172] *The Daily Telegraph*, 18 October 1995
[173] Dawkins (2006: p. 260)
[174] Graves (1929: p. 158)
[175] Cartwright (2000: p. 302)
[176] Wright (1971: p. 147)
[177] Huxley (1989: p. 30)
[178] Frazer (1924: p. 211)
[179] Bible: 2 Samuel, ch. 11
[180] Kennedy (1999)
[181] Ridley (1993: p. 233)
[182] Mill (1859)
[183] Eibl-Eibesfeldt (1996: pp. 164 *et seq.*) and Kennedy L (1999: p. 76)
[184] National Geographic Magazine, September 2007, p. 59
[185] Watson (2001)
[186] Shorter Oxford English Dictionary (1975)
[187] Hamilton (2001)
[188] Eibl-Eibesfeldt (1996)
[189] Hamilton (2001: p. 271)
[190] Encyclopaedia Britannica (1991: v. 16, p. 764)
[191] Encyclopaedia Britannica (1991: v. 19, p. 641)
[192] Darwin (1859: p. 236)
[193] Olivia Judson, *In Memory of Bill Hamilton*, http://evolution.unibas.ch/hamilton/economist.htm.
[194] Hamilton (1996: p. 24)

[195] Segerstråle (2013)
[196] Darwin (1860: p. 237)
[197] Hamilton (1996: p. xiv)
[198] *The Daily Telegraph*, 9 March 2000, Hamilton's obituary
[199] Lewis (2003: p. 129)
[200] Spencer (2006: p. 8)
[201] Bolitho (1929: p. 102); Ishaq throughout
[202] Andrae (2012: p. 14)
[203] Bolitho (1929: p. 101)
[204] Lings (1991: p. 1) and Bible: Genesis 16
[205] Lings (1991: p. 1) and Bible: Genesis 17.17
[206] Ishaq (1955: p. 45, para. 5)
[207] Ishaq (1955: p. 69, para. 5)
[208] Bolitho (1929: p. 102) and Ishaq (1955: p. 72, para. 2)
[209] Ishaq (1955: p. 73, para. 3)
[210] Ishaq (1955: p. 79, para. 5)
[211] Ishaq (1955: p. 82, para. 1)
[212] Bolitho (1929: p. 104)
[213] Ishaq (1955: p. 82, para. 2)
[214] Ishaq (1955: p. 83, para. 1)
[215] Ishaq (1955: p. 84, para. 1)
[216] Bolitho (1929: pp. 105 *et seq.*)
[217] Ishaq (1955: p. 107, para. 2)
[218] Ishaq (1955: p. 118)
[219] Ishaq (1955: p. 167, para. 1)
[220] Ishaq (1955: p. 165, para. 4; p. 166, para. 2)
[221] Ishaq (1955: p. 181; p. 182, para. 1 and footnote)
[222] Ishaq (1955: p. 183, para. 3)
[223] Spencer (2006: p. 89)
[224] Lewis (2003: p. 28)
[225] Ishaq (1955: p. 231, paras. 1 and 3)
[226] Ishaq (1955: pp. 235 *et seq.*)
[227] Ishaq (1955: p. 254, para. 3)
[228] Ishaq (1955: p. 267, para. 1; p. 684, para. 2)
[229] Ishaq (1955: pp. 242 *et seq.*; p. 254, para. 3; p. 266, para. 2)
[230] Ishaq (1955: p. 202, para. 2)
[231] Bolitho (1929: p. 104)
[232] Ishaq (1955: p. 222, para. 1)
[233] Ishaq (1955: p. 252 footnote)
[234] Spencer (2006: p. 30)
[235] Quran: Sura 48.27 and Spencer (2006: 20)
[236] Ishaq (1955: p. 493, para. 5)
[237] Ishaq (1955: p. 792, Ibn Hisham's note)
[238] Quran: Sura 66.1 to 5 and Spencer (2006: 21)
[239] Spencer (2006: 170)

[240] Ishaq (1955: p. 255, para. 2)
[241] Ishaq (1955: p. 516, para. 1)
[242] Spencer (2006: p. 5)
[243] Ishaq (1955: p. 286, para. 5)
[244] Ishaq (1955: p. 369, para. 4; p. 561) and Spencer (2006: 106, 116)
[245] Ishaq (1955: p. 466, para. 5; p. 493, para. 1; p. 509, para. 3; p. 511, para. 5; 515, para. 1)
[246] Ishaq (1955: p. 308, para. 5; p. 465, para. 1)
[247] Ishaq (1955: p. 665)
[248] Ishaq (1955: p. 287, para. 2; p. 308, para. 5; p. 438, para. 3; p. 516, para. 1; p. 523, para. 2; p. 592)
[249] Ishaq (1955: pp. 185 *et seq.*; p. 199, para. 5; p. 300, para. 3)
[250] Hourani (1991: 23), Lewis (2003: p. 32) and Spencer (2006: p. 98)
[251] Spencer (2006: p. 99)
[252] Spencer (2006: p. 146)
[253] Lewis (2003: p. 33)
[254] Ishaq (1955: p. 602, para. 2)
[255] Ishaq (1955: p. 682, para. 5) and Spencer (2006: p. 166)
[256] Hourani (1991: p. 22)
[257] Encyclopaedia Britannia (1991: v. 2, p. 699)
[258] Hourani (1991: p. 25)
[259] Hourani (1991: p. 28)
[260] Hourani (1991: p. 26)
[261] Hourani (1991: p. 26)
[262] Hourani (1991: p. 25)
[263] Spencer (2006: p. 152)
[264] Hourani (1991: p. 37)
[265] Hourani (1991: p. 31)
[266] Hourani (1991: p. 30 *et seq.*)
[267] Council on Foreign Relations. The Sunni–Shia divide. Available from: http://www.cfr.org/peace-conflict-and-human-rights/sunni-shia-divide/p33176#!/
[268] Reilly (2010: pp. 5 *et seq.*)
[269] Lewis (2003: p. 105) and Reilly (2010: pp. 43 *et seq.*)
[270] Lewis (2003: p. 99) and Reilly (2010: p. 164)
[271] Reilly (2010: p. 46)
[272] Reilly (2010: p. 76)
[273] Reilly (2010: p. 47)
[274] Reilly (2010: p. 52)
[275] Reilly (2010: p. 77)
[276] Lings (1991: p. 347) and Children of Prophet Muhammad (available from: http://islamicweb.com/history/children.htm)
[277] Lewis (2003: p. 111)
[278] Dowden (2008: p. 28)
[279] Lewis (2003: p. 4)

[280] Lewis (2003: p. 100)
[281] Wilkinson & Pickett (2009)
[282] Bowker (1998: p. 19 para 2), Horrie & Chippindale (1991: p. 25) and Quran 9.26
[283] Lewis (2003: pp. 96 *et seq.*)
[284] Lewis (2003: p. 115)
[285] Richardson (2013: p. 37)
[286] Mill (1859)
[287] Scruton (2000: ch. 6)
[288] Lewis (2003: p. 96)
[289] Lewis (2003: p. 146)
[290] Lewis (2003: p. 124)
[291] Masters (1954)
[292] Diamond (1991: p. 106)
[293] Buchan (1997: p. 34)
[294] Scruton (1995)
[295] 'How quickly do different cells in the body replace themselves?' Available from: http://www.weizmann.ac.il/plants/Milo/images/cellsBody ReplacementRate121118Clean(2).pdf
[296] *Medical Microbiology*, fourth edition, 1996. Chapter 95: Microbiology of the Gastrointestinal Tract. Available from: http://www.ncbi.nlm.nih.gov/books/NBK7670/
[297] Eliot (2001) and Tancredi (2005)
[298] Hansen (2009: p. 100)
[299] Mishan (1993: p. 47)
[300] Odum (1954: p. 134)
[301] Wikipedia article on *Irrigation statistics*. Available from: http://en.wikipedia.org/wiki/Irrigation_statistics
[302] Russell (1950: p. 452)
[303] Flannery (1994: p. 93) and Grove & Rackman (2001: pp. 62 *et seq.*)
[304] Dregne (1983: ch. 2)
[305] Diamond (1991: p. 297) and Encyclopaedia Britannica (1991: v. 3, p. 50)
[306] Diamond (1991: p. 294)
[307] Leakey & Lewin (1995: pp. 188 *et seq.*)
[308] Diamond (1991: p. 301) and Encyclopaedia Britannica (1991: v. 9, pp. 307 and 339)
[309] Spindler (1994: p. 251 *et seq.*)
[310] Andrews (1991: p. 64) and Spindler (1994: p. 242)
[311] Buchan (1997: ch 1, p. 7)
[312] Buchan (1997: p. 23)
[313] Buchan (1997: p. 19)
[314] Bible: 1 Timothy 6.10
[315] Wikipedia article on *Coin*. Available from: http://en.wikipedia.org/wiki/Coin#First_coins
[316] Buchan (1997: p. 108)

[317] *The Daily Telegraph*, 15 July 2015
[318] Bible: Exodus 22.25, Deuteronomy 20.19 to 20 and others
[319] Wikipedia article on *Usury*. Available from: http://en.wikipedia.org/wiki/Usury
[320] Thompson Seton (1951: pp. 9, 122, 304)
[321] Aristotle 4th Century BCE, 1.9 1257a, quoted by Buchan (1997: p. 23)
[322] Wikipedia article on *Coin* (http://en.wikipedia.org/wiki/Coin#First_coins)
[323] Wikipedia article on *Coin* (http://en.wikipedia.org/wiki/Coin#First_coins)
[324] Wikipedia article on *Coin* (http://en.wikipedia.org/wiki/Coin#First_coins)
[325] Maxwell (1966: p. 123)
[326] Mishan (1993: p. 161)
[327] Buarque (1993: p. 114), Buchan (2006) and Encyclopaedia Britannica (1991: v. 27, p. 311)
[328] Buarque (1993 pp. 12, 114), Encyclopaedia Britannica (1991: v. 27, p. 310, 392), Hardin (1968: p. 124)
[329] Smith (2000: p. 292)
[330] Buchan(2006: p. 2)
[331] Buchan (2006: p. 28)
[332] Buarque (1993: pp. 21, 73, 104, 112)
[333] Buarque (1993: p. 70)
[334] Bible: Mark 5.1 to 13. See also About Religion (available from: http://atheism.about.com/od/biblegospelofmark/a/mark05a.htm)
[335] Buchan (1997: p. 179)
[336] Wikipedia article on *Carbon credit*. Available from: http://en.wikipedia.org/wiki/Carbon_credit
[337] Hansen (2009: p. 83)
[338] Hansen (2009: p. 160)
[339] Wilkinson & Pickett (2011)
[340] Wilson E (1984)
[341] Cramp (1980: v. 2, p. 419 onwards)
[342] Hamilton (2001: p. 798), also see index
[343] *Nature*, 28 October 1982, v. 299, pp. 818 to 820
[344] Berlin (1959: p. 217)
[345] Hamilton (2001: p. 463)
[346] Hamilton (2001: ch. 12) and almost any newspaper of the second decade of the 21st Century
[347] Berlin (2008: p. 210)
[348] Mill (1859: p. 97)
[349] http://www.bbc.co.uk/news/health-34208624
[350] Mill (1859: p. 67)
[351] The World Factbook – CIA. Long-term global demographic trends; reshaping the geopolitical landscape. Available from: https://www.cia.gov/library/reports/general-reports-1/Demo_Trends_For_Web.pdf
[352] Cain (1993b: chs. 6 and 7)

[353] BBC News. Not one but 'six giraffe species'. Available from: http://news.bbc.co.uk/2/hi/sci/tech/7156146.stm
[354] Burke (1790) and Scruton (2006: p. 176 *et seq.*)
[355] Dawkins (1989: p. 215)
[356] Crowder (1962)
[357] O'Brian (1989: p. 256)
[358] Mill (1859: p. 94)
[359] Sen (2009: pp. xvii *et seq.*)
[360] de Waal (1998)
[361] Morris (1971: ch. 4)
[362] Sykes (2001)
[363] Dudley Stamp (1969: pp. 36 *et seq.*)
[364] Stenton (1947: pp. 274 *et seq.*)
[365] Stenton (1947: p. 284)
[366] Oxford English Dictionary Compact Edition (1979: v. 1, p. 963)
[367] Stenton (1947: p. 497)
[368] Stenton (1947: p. 186)
[369] Stenton (1947: p. 643)
[370] Baily, Gunn & Smith (1991: p. 5)
[371] Scruton (2006)
[372] Garmonsway (ed.) (1960: p. 200) Anglo Saxon Chronicle
[373] Baily, Smith & Gunn (1991: pp. 5, 31)
[374] Winston Churchill, *Hansard*, 11 November 1947
[375] Dunbar (1996), especially p. 79
[376] Mill (1859: p. 101)
[377] Paradis (1989: p. 30)
[378] Jet Propulsion Laboratory. California Institute of Technology. Where are the Voyagers? Voyager 1. Available from: http://voyager.jpl.nasa.gov/where/index.html
[379] Skidelsky (2002: p. 116)
[380] Darwin (1845/1902: pp. 212 to 213)
[381] Sobel (1998: p. 164)
[382] Williams (1953: pp. 206 to 210)
[383] Scruton (2010)

# References

ABERCROMBIE, M. (1960) *The Anatomy of Judgement: an Investigation into the Processes of Perception and Reasoning,* London: Hutchinson.

ALI, A. (trans. 1934) *The Holy Quran*, Islamic Propagation Centre International, (1946).

ALLMAN, W. (1994) *The Stone Age Present: How Evolution Has Shaped Modern Life. From Sex, Violence, and Language to Emotions, Morals, and Communities,* New York, NY: Simon & Schuster.

ANDERSON, P. (1971) *Omega: Murder of the Ecosystem and Suicide of Man,* Dubuque, IA: W. C. Brown Co.

ANDRAE, T. (2012) *Mohammed: the Man and His Faith,* Dover Publications.

ANDREWARTHA, H. & BIRCH, C. (1984) *The Ecological Web: More on the distribution and Abundance of Animals,* Chicago, IL: University of Chicago Press.

ANDREWS, M. (1991) *The Birth of Europe: Colliding Continents and the Destiny of Nations,* London: BBC Books, a division of BBC Enterprises.

AXELROD, R. (1984) *The Evolution of Cooperation,* Toronto: Penguin Books.

BAILEY, S., GUNN, M. & SMITH, P. (1991) *Smith and Bailey on the Modern English Legal System,* London: Sweet & Maxwell.

BARRETT, P. & PALMER, D (2009) *Evolution: The Story of Life,* London, Mitchell Beazley.

BERLIN, I. (1968) *Liberty,* Oxford: OUP (2008).

BERNE, E. (1974) *What Do You Say After You Say Hello? The Psychology of Human Destiny,* Corgi.

[THE HOLY] BIBLE CONTAINING THE OLD AND NEW TESTAMENTS (AUTHORIZED KING JAMES VERSION) (1957). London: Collins.

BILLINGS, W. (1964) *Plants and the Ecosystem,* London: Macmillan.

BIRKHEAD, T. (2000) *Promiscuity: an Evolutionary History of Sperm Competition,* London: Faber.

BLACKBURN, S. (1998) *Ruling Passions: A Theory of Practical Reasoning,* Oxford: Clarendon Press, (2000).

BLACKBURN, S. (1999) *Think: a Compelling Introduction to Philosophy,* Oxford: OUP.

BLACKMORE, S. (1999) *The Meme Machine,* Oxford: OUP.

BLUNT, W. (1971) *The Compleat Naturalist: a Life of Linnaeus,* London: Collins.

BODANIS, D. (2000) *$E=mc^2$: a Biography of the World's Most Famous Equation,* London: Macmillan.

BOLITHO, W. (1929) *Twelve Against the Gods: the Story of Adventure,* New York, NY: Simon and Schuster.

BOWKER, J. (1998) *What Muslims Believe,* Oxford: Oneworld.

BOWLBY, J. (1990) *Charles Darwin: a New Life,* London: Hutchison.

BROWN, E. (1995) *Charles Darwin Voyaging,* Princeton, NJ: Princeton University Press.

BROWN, E. J. (2002) *Charles Darwin: The Power of Place,* Princeton, NJ: Princeton University Press.

BUARQUE, C. (1993) *The End of Economics? Ethics and the Disorder of Progress,* Atlantic Highlands, NJ: Zed Books.

BUCHAN, J. (1997) *Frozen Desire: an Inquiry into the Meaning of Money,* London: Picador.

BUCHAN, J. (2004) *Capital of the Mind: How Edinburgh Changed the World,* London: J. Murray.

BUCHAN, J. (2006) *Adam Smith and the Pursuit of Perfect Liberty,* London: Profile.

BURKE, E. (1790) *Reflections on the Revolution in France,* London: J. Dodsley.

BYRNE, R. & WHITEN, A. (eds.) (1988) *Machiavellian Intelligence: Social Expertise and the Evolution of Intellect in Monkeys, Apes, and Humans,* Oxford: OUP.

CAIN, A. (1993a) Letter to the Editor. *The Linnean,* **9**: 3.

CAIN, A. (1993b) *Animal Species and Their Evolution,* Princeton, NJ: Princeton University Press.

CALDER, N. (1972) *The Restless Earth: a Report on the New Geology,* London: BBC.

CARTWRIGHT, J. (2000) *Evolution and Human Behaviour: Darwinian Perspectives on Human Nature,*

Basingstoke: Macmillan.
CHERRETT, J. & BRADSHAW, A. (1989) *Ecological Concepts: the Contribution of Ecology to an Understanding of the Natural World,* Oxford: Blackwell Scientific Publication.
CLUTTON-BROCK, J. (1999) *A Natural History of Domesticated Mammals,* Cambridge: CUP.
COHEN, J. (1995) *How Many People Can the Earth Support?* New York, NY: Norton.
CRAMP, S. (ed.) (1980) *Handbook of the Birds of Europe, the Middle East and North Africa: the Birds of the Western Palearctic,* Oxford: OUP.
CROWCROFT, P. (1991) *Elton's Ecologists: a History of the Bureau of Animal Population,* Chicago, IL: University of Chicago Press.
CROWDER, M. (1962) *The Story of Nigeria,* London: Faber and Faber.
DALEY, H., COBB, J. & COBB, C. (1994) *For the Common Good: Redirecting the Economy Toward Community, the Environment, and a Sustainable Future,* Boston, MA: Beacon Press.
DARWIN, C. (1860) *On the Origin of Species,* London: Murray.
DARWIN, C. (1845) *A Naturalist's Voyage Round the World: the Voyage of the Beagle,* London: John Murray (1902).
DARWIN, F. (ed. 1892) *Autobiography of Charles Darwin and Selected Letters,* New York, NY: Dover, (1958).
DAWKINS, M., HALLIDAY, T. & DAWKINS, R. (eds.) (1991) *The Tinbergen Legacy,* London: Chapman & Hall.
DAWKINS, M. (1993) *Through Our Eyes Only? The Search for Animal Consciousness,* Oxford: W. H. Freeman.
DAWKINS, R. (1989) *The Selfish Gene,* Oxford: OUP.
DAWKINS, R. (2004) *The Ancestor's Tale,* Boston, MA: Houghton Mifflin.
DAWKINS, R. (2006) *The God Delusion,* London: Bantam Press.
DAWOOD, N. (1956) *The Koran,* Harmondsworth: Penguin Books.

DE MONTAIGNE, M. (1580) *Four Essays* London: Penguin (1995).

DE WAAL, F. (1996) *Good Natured: the Origins of Right and Wrong in Humans and Other Animals,* Cambridge, MA: Harvard University Press.

DE WAAL, F. (1998) *Chimpanzee Politics: Power and Sex Among Apes,* Baltimore, MD: Johns Hopkins University Press.

DENNETT, D. (1995) *Darwin's Dangerous Idea: Evolution and the Meanings of Life,* London: Penguin.

DENNETT, D. (2006) *Breaking the Spell: Religion as a Natural Phenomenon,* London: Penguin.

DESMOND, A. & MOORE, J. (1992) *Darwin*. London: Penguin.

DIAMOND, J. (1991) *The Rise and Fall of the Third Chimpanzee,* London: Vintage.

DIAMOND, J. (1998) *Guns Germs and Steel,* London: Vintage.

DIAMOND, J. (2000) *Why is Sex Fun? The Evolution of Human Sexuality,* London: Phoenix.

DICKINSON, G. & MURPHY, K. (1998) *Ecosystems: a Functional Approach,* London: Routledge.

DONOVAN, S. (1989) *Mass Extinctions,* London: Belhaven Press.

DOWDEN, R. (2008) *Africa: Altered States, Ordinary Miracles,* London: Portobello.

DREGNE, H. (1983) *Desertification of Arid Lands,* Harwood Academic Publishers.

DUDLEY STAMP, L. (1969) *Man and the Land,* London: Collins.

DUNBAR, R. (1995) *The Trouble with Science,* London: Faber.

DUNBAR, R. (1996) *Grooming, Gossip, and the Evolution of Language,* London: Faber and Faber.

DUNBAR, R. (2010) *How Many Friends Does One Person Need? Dunbar's Number and Other Evolutionary Quirks,* London: Faber and Faber.

EHRLICH, P. (1971) *The Population Bomb,* New York, NY: Ballantine Books.

EIBL-EIBESFELDT, I. (1996) *Love and Hate: the Natural History of Behavior Patterns,* Aldine de Gruyter.

ELIOT, L. (1999) *Early Intelligence,* London: Penguin.
ELTON, C. (1953) *The Ecology of Animals,* London: Methuen.
FLANNERY, T. (1994) *The Future Eaters: an Ecological History of the Australasian Lands and People,* Sydney: New Holland Publishers.
FLANNERY, T. (2010) *Here on Earth: a New Beginning*, Allen Lane.
FRAZER, J. & FRAZER, R. (1924) *The Golden Bough: a Study in Magic and Religion,* London: Macmillan.
GARMONSWAY, G. N. (ed.) (1960) *Anglo Saxon Chronicle*, London: Dent.
GOODALL, J. (1971) *In the Shadow of Man,* Boston, MA: Houghton Mifflin.
GOODALL, J. (1990) *Through a Window: My Thirty Years with the Chimpanzees of Gombe,* Boston, MA: Houghton Mifflin.
GORDON, R. (1990) *The Structure of Emotions: Investigations in Cognitive Philosophy,* Cambridge: CUP.
GOULD, S. (1977) *Ever Since Darwin: Reflections in Natural History,* New York, NY: Norton.
GOULD, S. (1980) *The Panda's Thumb: More Reflections in Natural History,* New York, NY: Norton.
GOULD, S. (2000) *The Lying Stones of Marrakech: Penultimate Reflections in Natural History,* London: Jonathan Cape.
GRAVES, R. (1929) *Good-Bye to All That,* Harmondsworth: Penguin Books.
GRAVES, R. (1955) *The Greek Myths,* Baltimore, MD: Penguin Books.
GREIG-SMITH, P. (1983) *Quantitative Plant Ecology* Oxford, Blackwell Scientific Publications.
GROVE, A. & RACKHAM, O. (2001) *The Nature of Mediterranean Europe: an Ecological History,* New Haven, CT: Yale University Press.
HAMILTON, W. (1995) *Narrow Roads of Gene Land: the Collected Papers of W. D. Hamilton,* New York, NY: W. H. Freeman.
HAMILTON, W. (2001) *Narrow Roads of Gene Land: the Collected Papers of W. D. Hamilton,* Oxford: OUP.

HANSEN, J. (2009) *Storms of my Grandchildren: the Truth About the Coming Climate Catastrophe and Our Last Chance to Save Humanity,* New York, NY: Bloomsbury USA.

HARARI, Y. (2011) *Sapiens: a Brief History of Mankind,* London: Vintage Books.

HARDIN, G. (1968) The tragedy of the commons, *Science,* **162**: 1243–1248.

HARRIS, T. (1970) *I'm OK, You're OK: a Practical Guide to Transactional Analysis,* London: Cape.

HAWKING, S. (2001) *The Universe in a Nutshell,* New York, NY: Bantam Books.

HOBBES, T. (1986) *Leviathan,* Harmondsworth: Penguin Classics.

HORRIE, C. & CHIPPINDALE, P. (1991) *What is Islam? A Comprehensive Introduction,* London: Virgin.

HOURANI, A. (1991) *A History of the Arab Peoples,* London: Faber and Faber.

HUME, D. (1709) *A Treatise of Human Nature,* London: Penguin Books, (1984).

HUXLEY, T., PARADIS, J. & WILLIAMS, G. (1989) *Evolution and Ethics: T.H. Huxley's "Evolution and Ethics" with New Essays on its Victorian and Sociobiological Context,* Princeton, NJ: Princeton University Press.

ISHAQ, I. (1st Century AH) *The Life of Muhammad: a Translation of Ishaq's Sirat Rasul Allah,* London: OUP, (1955).

KANT, I. (1785) *The Moral Law: Groundwork of the Metaphysic of Morals,* London: Routledge, (2010).

KEEGAN, J. (1994) *A History of Warfare,* London: Pimlico.

KENNEDY, L. (1999) *All in the Mind: A Farewell to God,* London: Sceptre.

KENNEDY, P. (1994) *Preparing for the Twenty-first Century*, London: Fontana Press.

KEYNES, J. (1923) *A Tract on Monetary Reform,* B N Publishing, (2008).

KORMONDY, E. (1969) *Concepts of Ecology,* Englewood Cliffs, NJ: Prentice-Hall.

KRUUK, H. (2003) *Niko's Nature: the Life of Niko Tinbergen and His Science of Animal Behaviour,* Oxford: OUP.
LAMBERT, J. (1967) *The Teaching of Ecology: a Symposium of the British Ecological Society, Goldsmiths College, University of London, UK, 13 April–16 April 1966,* Oxford: Blackwell Scientific.
LEAKY, R. & LEWIN, R. (1992) *Origins Reconsidered: in Search of What Makes Us Human,* London: Little Brown.
LEAKY, R. & LEWIN, R. (1995) *The Sixth Extinction,* London: Phoenix.
LEVITIN, D. (2015) *The Organised Mind: Thinking Straight in the Age of Information Overload,* London: Viking.
LEWIS, B. (2003) *The Crisis of Islam: Holy War and Unholy Terror,* London: Phoenix.
LEWIS, M. (1989) *Liar's Poker,* London: Hodder and Stoughton.
LEWIS-WILLIAMS, D. (2004) *The Mind in the Cave: Consciousness and the Origins of Art,* London: Thames & Hudson.
LINGS, M. (1991) *Muhammad: His Life Based on the Earliest Sources,* London: Islamic Texts Society.
LINK, P. (2011) China: from famine to Oslo, *New York Review of Books*, 13 January 2011.
LIVINGSTONE SMITH, D. (2007) *The Most Dangerous Animal: Human Nature and the Origins of War,* New York, NY: St Martin's Press.
LOMBORG, B. (2001) *The Skeptical Environmentalist: Measuring the Real State of the World,* Cambridge: CUP.
MAALOUF, A. (1997) *The Gardens of Light,* Abacus.
MAGEE, B. (1985) *Popper,* London: Fontana Press.
MALTHUS, T. (1960) *On Population: Three Essays,* New York, NY: New American Library.
MAREAN, C. (2010) When the sea saved humanity, *Scientific American*, 303: 54–61.
MASTERS, J. (2012) *Bhowani Junction,* Souvenir Press (Kindle Edition).
MATTHIESSEN, P. (1965) *At Play in the Fields of the Lord,*

New York, NY: Random House.
MAXWELL, G. (1966) *Lords of the Atlas: the Rise and Fall of the House of Glaoua, 1893–1956,* London: Arrow.
MAYNARD SMITH, J. & SZATHMÁRY, E. (1999) *The Origins of Life: from the Birth of Life to the Origin of Language,* Oxford: OUP.
MILL, J. (1859) *On Liberty,* New York, NY: Cosimo Classics (2006).
MILLER, J. (1986) *Norman Angell and the Futility of War: Peace and the Public Mind,* London: Macmillan.
MISHAN, E. (1993) *The Costs of Economic Growth,* London: Weidenfeld and Nicolson.
MITHEN, S. (1996) *The Prehistory of the Mind: a Search for the Origins of Art, Religion, and Science,* London: Phoenix.
MITTERMEIER, R., RYLANDS, A., WILSON, D., ANANDAM, M. & BRAULIK, G. (eds.) (2013) *Primates,* Barcelona: Lynx.
MONBIOT, G. (1994) *No Man's Land: An Investigative Journey Through Kenya and Tanzania,* London: Picador.
MORRIS, D. (1971) *The Human Zoo,* London: World Books.
MURRAY, A. (1994) *Who's Who in Mythology: A Classic Guide to the Ancient World,* London: Bracken Books.
NEAL, E. (1953) *Woodland Ecology,* London: Heinemann.
NICHOLS, D. (1962) *Echinoderms,* London: Hutchinson University Library.
O'BRIAN, P. (1989) *Joseph Banks: a Life,* London: Collins Harvill.
O'HEAR, A. (1999) *Beyond Evolution: Human Nature and the Limits of Evolutionary Explanation,* Oxford: OUP.
ODUM, E. (1954) *Fundamentals of Ecology,* Philadelphia, PA: Saunders.
ODUM, E. (1963) *Ecology,* New York, NY: Holt, Rinehart and Winston.
PINKER, S. (1994) *The Language Instinct,* London: Penguin Books.
PINKER, S. (1997) *How the Mind Works,* London: Penguin Press.

POLUNIN, N. (ed.) (1998) *Population and Global Security,* Cambridge: CUP.
POPPER, K. (1959) *The Logic of Scientific Discovery,* London: Routledge, (2002).
PROSS, A. (2012) *What is Life? How Chemistry Becomes Biology,* Oxford: OUP.
REILLY, R. (2010) *The Closing of the Muslim Mind: How Intellectual Suicide Created the Modern Islamist Crisis,* Wilmington, DE: ISI Books.
RICHARDSON, H. (2013) *The Story of Mohammed: Islam Unveiled,* North Charleston, SC: CreateSpace Independent Publishing Platform.
RIDLEY, M. (1993) *The Red Queen: Sex and the Evolution of Human Nature,* London: Viking.
RIDLEY, M. (1996) *The Origins of Virtue: Human Instincts and the Evolution of Cooperation,* London: Viking.
RIDLEY, M. (2010) *The Rational Optimist: How Prosperity Evolves,* New York, NY: Harper.
ROMER, A. (1933) *Man and the Vertebrates,* Harmondsworth: Penguin Books.
ROUSSEAU, J.-J. (1762) *The Social Contract,* Ware: Wordsworth Editions Ltd., (1998).
ROYAL SOCIETY OF GREAT BRITAIN. (2012) *People and the Planet,* London: The Royal Society.
RUSSELL, E. (1950) *Soil Conditions & Plant Growth,* London: Longmans, Green.
SAYERS, D. (1930) *The Documents in the Case,* London: Ernest Benn.
SCRUTON, R. (1982) *Kant: A Very Short Introduction,* Oxford: OUP.
SCRUTON, R. (1995) *A Short History of Modern Philosophy: from Descartes to Wittgenstein,* London: Routledge.
SCRUTON, R. (2000) *England: an Elegy,* London: Chatto & Windus.
SCRUTON, R. (2002) *Spinoza: A Very Short Introduction,* Oxford: OUP.
SCRUTON, R. (2005) *Modern Culture,* London: Continuum.
SCRUTON, R. (2006) *A Political Philosophy,* London: Continuum.

SCRUTON, R. (2010) *The Uses of Pessimism: and the Danger of False Hope,* London: Atlantic Books.
SEGERSTRÅLE, U. (2013) *Nature's Oracle: The Life and Work of W. D. Hamilton,* Oxford: OUP.
SEN, A. (2009) *The Idea of Justice,* London: Allen Lane.
SHUTE, N. (1953) *In the Wet,* New York, NY: Morrow.
SKIDELSKY, R. (2002) *John Maynard Keynes,* London: Penguin.
SMITH, A (1776) *An Inquiry into the Nature and Causes of the Wealth of Nations,* Oxford: University Press, (2008).
SOBEL, D. (1998) *Longitude,* London: Fourth Estate.
SOBER, E. & WILSON, D. (1998) *Unto Others,* Cambridge, MA: Harvard University Press.
SOLOMON, M. (1969) *Population Dynamics,* London: Edward Arnold.
SPENCER, R. (2006) *The Truth About Muhammad: Founder of the World's Most Intolerant Religion,* Washington, DC: Regnery Publishing.
SPINDLER, K. (1994) *The Man in the Ice: the Preserved Body of a Neolithic Man Reveals the Secrets of the Stone Age,* London: Weidenfeld and Nicolson.
STAMP DAWKINS, M. (1993) *Through Our Eyes Only? The Search for Animal Consciousness,* Oxford: W. H. Freeman.
STENTON, F. (1947) *Anglo-Saxon England,* Oxford: Clarendon Press.
SYKES, B. (2001) *The Seven Daughters of Eve: the Science that Reveals Our Genetic Ancestry,* New York, NY: Norton.
TANCREDI, L. (2005) *Hardwired Behaviour: What Neuroscience Reveals about Morality,* Cambridge: CUP.
TAYLOR, J. (2012) *Petra and the Lost Kingdom of the Nabataeans*, London: I. B. Tauris
TEMPLETON, A. (2002) Out of Africa again and again. *Nature*, **416**: 45–51.
TERBORGH, J. (2007) Book review of *The Last Forest: the Amazon in the Age of Globalisation,* Random House. *New York Review of Books,* 22 November 2007.

THE NEW ENCYCLOPAEDIA BRITANNICA (1991) London: Encyclopaedia Britannica.

THOMPSON SETON, E. (1951) *Trail of an Artist-Naturalist: an Autobiography of Ernest Thompson Seton,* London: Hodder & Stoughton.

TINBERGEN, N. (1951) *The Study of Instinct,* Oxford: Clarendon Press.

UCKO, P. & DIMBLEBY, G. (eds.) (1969) *The Domestication and Exploitation of Plants and Animals,* London: Duckworth.

WALLACE, A. (1869) *The Malay Archipelago: The Land of the Orang-Utan, and the Bird of Paradise. A Narrative of Travel, with Studies of Man and Nature,* London: Macmillan.

WARBURTON, N. (1992) *Philosophy: the Basics* London: Routledge.

WATSON, P. (2001) The terrible truth about paradise. *Financial Times Weekend,* 23 December 2001.

WILKINSON, R. & PICKETT, K. (2009) *The Spirit Level: Why Greater Equality Makes Societies Stronger,* New York, NY: Bloomsbury Press.

WILLIAMS, G. (1992) *Natural Selection: Domains, Levels, and Challenges,* New York: OUP.

WILLIAMS, J. (1953) *Bandoola,* London: Hart-Davis.

WILLIAMSON, M. (1992) Haldane's special preference. *The Linnean,* **8**: 14.

WILSON, A. (1997) *Paul: the Mind of the Apostle,* London: Sinclair-Stevenson.

WILSON, A. (2003) *The Victorians,* London: Arrow Books.

WILSON, E. (1984) *Biophilia,* Cambridge, MA: Harvard University Press.

WILSON, E. (1994) *Naturalist,* London: Allen Lane.

WILSON, E. (1998) *Consilience: the Unity of Knowledge,* London: Abacus.

WRIGHT, D. (1971) *The Psychology of Moral Behaviour,* Harmondsworth: Penguin Books.

# INDEX

Printed in Great Britain
by Amazon.co.uk, Ltd.,
Marston Gate.